Lecture Notes in Mathematics

Edited by A. Dold and B. Eckmann

T0216098

733

Frederick Bloom

Modern Differential Geometric Techniques in the Theory of Continuous Distributions of Dislocations

Springer-Verlag
Berlin Heidelberg New York 1979

Author

Frederick Bloom
Department of Mathematics,
Computer Science and Statistics
University of South Carolina
Columbia, S.C. 29208
USA

AMS Subject Classifications (1970): Primary: 73S05
Secondary: 53C10

ISBN 3-540-09528-4 Springer-Verlag Berlin Heidelberg New York
ISBN 0-387-09528-4 Springer-Verlag New York Heidelberg Berlin

Library of Congress Cataloging in Publication Data
Bloom, Frederick, 1944 –
Modern differential geometric techniques in the
theory of continuous distributions of dislocations.
(Lecture notes in mathematics ; 733)
Bbiliography: p.
Includes index.
1. Dislocations in crystals. 2. Geometry, Differential. 3. G-structures. I. Title.
II. Series: Lecture notes in mathematics (Berlin) ; 733.
OD921.B56 548'.842 79-9374
ISBN 0-387-09528-4

Printing and binding: Beltz Offsetdruck, Hemsbach/Bergstr.
2141/3140-543210

FOR

HARRY

AND

MORDECHAI

Preface

Among research workers in mechanics and applied mathematics there has been great interest, in the past two decades, in the area of continuum theories of dislocations. More recently, attention has turned to the more difficult problem connected with the motion of dislocations through a continuum and its relation to various formulations of plasticity theory.

The first comprehensive effort to formulate a geometric theory for a body possessing a continuous distribution of dislocations was made by Kondo ([1],[2]) and consisted in prescribing, on the basis of certain heuristic arguments, various geometric structures on the body manifold, such as a metric and an affine connection, which then served to characterize certain properties of the dislocation distribution; similar efforts were made by Bilby and by Kröner ([3], [4]), however, in none these theories was any serious consideration given to the types of constitutive equations which may be associated with the body manifold.

A new approach to the problem was made by Noll [5] in the early sixties and was later extended by Noll [6] and by Wang [7]. Here one starts with the prescription of a constitutive equation for particles belonging to the body manifold and, using the concept of a uniform reference, develops a geometric theory which in many ways is isomorphic to those considered by Kondo, Bilby, and Kröner. Wang's work differs from that of Noll in that the body need only admit a uniform reference locally; their work represents the first known use of concepts belonging to the realm of modern differential geometry in formulating a theory in continuum mechanics.

An application of the concepts developed by Noll and Wang has been made by Toupin [8] (to a theory of dislocations in crystalline media) and Bilby [9] and Kröner [10] have both examined, in detail, the relationships which exist among the approaches taken by Noll and Wang, on the one hand, and by themselves and Kondo, on the other; due to considerations of space we will have to ignore such developments here just as we shall pass over recent work of Wang, et. al, on wave behavior in inhomogeneous elastic bodies [11], [12], [13], and on classes of universal solutions [14].

A theory of dislocation motion in a continuum was formulated by Eckart [15] in 1948 in a proposal he dubbed "anelasticity". Eckart suggested as a model for a body containing a continuous distribution of dislocations, which may be moving within the body manifold, a continuum in which the Cauchy stress arises in response to deformations from natural states which may be different, for different particles, and, perhaps, also varying with time. Eckart's proposal was examined by Truesdell in [16] and by Truesdell and Toupin in [17] and in attempt was then made by Bloom [18] to fi the basic tenets of Eckart's proposal into the framework developed by Wang for static dislocation distributions; a correct formulation of Eckart's anelasticity proposal, within this differential geometric setting, was given by Wang and Bloom [19] and has been extended by them in [20] to allow for thermodynamic influences. More recently Wang [39] has sought to formulate the connection between anelastic response and recent ideas concerning materials with elastic range.

Our aim in preparing this monograph has been not only to try to present an accurate picture of the current status of dislocation theory, as a branch of continuum mechanics, but also to illustrate an important application of modern differential geometric ideas in physics. This is the proper place to acknowledge a debt of gratitude to Professor C. C. Wang who has been the outstanding major contributor to this important new area of continum physics. Finally the author would like to thank Mrs. Margaret Robinson, for the excellent job of typing she has done, and the college of Science and Mathematics at USC for a grant during the summer of 1975 which enabled me to complete the greater part of the work presented here.

TABLE OF CONTENTS

Chapter I. Mathematical Preliminaries

1. Introduction

We wish to outline here those elements of differential geometry with which the reader should be conversant in order to understand the text. As in past volumes in this series, we shall assume that the reader is familiar with those basic concepts which underlie the differential manifold approach to differential geometry and the theory of Lie groups. Thus we aim, essentially, at setting the notation which we shall use in what follows.

The approach to manifold theory which has been employed in most of the recent continuum mechanics literature on dislocation theory is that of Kobayashi and Nomizu and the main reference here would be [21]. Alternatively, the reader may consult the excellent exposition of differential geometry that is to be found in [22] and [23]; these later volumes have strongly influenced the author's viewpoint of manifold theory and we shall rely on them as we present the definitions and theorems below.

2. Differentiable Manifolds

Definition I-1. A differentiable manifold of class k and dimension n is a pair consisting of a Hausdorff space M with a countable base and a set F of real valued functions which are defined on open sets of M and which have the following properties:

(i) if $f \in F$ is defined on U (an open set in M) and V is open in U then $f|_V$ is in F; if f is defined on U (open set in M) where $U = \bigcup_{\alpha \in I} U_\alpha$ (U_α, $\alpha \in I$, open in M) then $f \in F$ if $f|_{U_\alpha}$ is in F for each $\alpha \in I$.

(ii) for each $p \in M$, there exists an open neighborhood U containing p and a homeomorphism ϕ: $U \to \phi(U) \subset R^n$ such that if V is open in U, the set of all $f \in F$ which are defined on V is identical with $C^k(\phi(V)) \circ \phi$.

The functions $f \in F$ are called differentiable functions and the Hausdorff space M is the underlying space of the manifold; an open set U containing $p \in M$ which satisfies (ii) above is called a coordinate neighborhood of p and the homeomorphism ϕ is called a coordinate map near p. If $q \in U$ then $\phi(q) = (x^1(q), \ldots x^n(q))$ where the $x^i(q)$, $i = 1,\ldots,n$ are the coordinate functions of ϕ (or the local coordinates of ϕ). The pair (U,ϕ) is a coordinate chart. A collection of coordinate charts $\{(U_\alpha, \phi_\alpha), \alpha \in I\}$, where $\{U_\alpha, \alpha \in I\}$ is an open covering of M, is called an atlas and it can be proven that an atlas completely determines a differentiable manifold if the maps

$$\kappa_{\alpha\beta} \equiv (\phi_\alpha \circ \phi_\beta^{-1})|_{\phi_\beta(U_\alpha \cap U_\beta)}, \quad \alpha,\beta \in I$$

are diffeomorphisms of class k.

Now, let M be a differentiable manifold of class k and let U be an open set in M. Following Wang [22] we denote by $F^*(k,U)$ the set of all differentiable functions defined on U and by $F^*(k,p)$, $p \in U$, the set of all equivalence classes of differentiable functions whose domains are open neighborhoods of p. If $f \in F^*(k,p)$ then f is said to be defined near p; if $f,g \in F^*(k,p)$ then there is some open neighborhood of p on which f and g agree. If (U,ϕ) is a chart for $p \in M$ then ϕ induces an isomorphism i: $F^*(k,p) \to F^*(k,\phi(p))$, i.e., $i(f) = f \circ \phi^{-1}$ (the representation of f via ϕ). Let $Z^*(k,\phi(p))$ be the subspace of $F^*(k,\phi(p))$ consisting of all differentiable functions defined near $\phi(p)$ which have zero first-order partial derivatives at $\phi(p)$. Under the isomorphism i, $Z^*(k,\phi(p))$ corresponds to a subspace $Z^*(k,p)$ of $F^*(k,p)$ in such a way that $f \in Z^*(k,p)$ iff $f \circ \phi^{-1}$ has zero first order partial derivatives at $\phi(p)$ and it is easy to show that $Z^*(k,p)$ is invariant under coordinate transformations near p, i.e., $Z^*(k,p)$ is independent of the chart (U,ϕ).

The subspace $Z^*(k,p)$ gives rise to the quotient space $M_p^* \equiv F^*(k,p)/Z^*(k,p)$ which we call the <u>cotangent space</u> of M at p. Let d_p, $p \in M$, denote the natural projection (a linear map) from $F^*(k,p)$ into $F^*(k,p)/Z^*(k,p)$. If f is a differentiable function in $F^*(k,p)$ then d_pf, its <u>differential</u>, is a <u>cotangent vector</u> in M_p^*. If (U,ϕ) is a chart of $p\in M$ and $f\in F^*(k,p)$ then, as the coordinate functions $x^i(p)$, $i = 1,\ldots,n$, are in $F^*(k,p)$ it is easy to see that $d_pf = \frac{\partial(f\circ\phi^{-1})}{\partial x^i}\Big|_p dx^i$ so that the set $\{d_px^i,\ i = 1,\ldots n\}$ forms a basis for M_p^*. The space dual to M_p^* is denoted by M_p and is called the <u>tangent space</u> to M at p; its elements are called <u>tangent vectors</u> and relative to the chart (U,ϕ) the basis in M_p which is dual in M_p^* is denoted by $\{\frac{\partial}{\partial x^i},\ i = 1,2,\ldots n\}$. If M is a manifold and $p\in M$, then M_p is susceptible of a rather concrete interpretation, as follows: Let a and b be real numbers and let $\lambda: (a,b) \to M$ such that $\lambda(c) = p$ for some $c\in(a,b)$. Let (U,ϕ) be a chart of p so that $\phi(\lambda(t)) \equiv (\lambda^1(t),\ldots,\lambda^n(t))$ for $a \le t \le b$. If the $\lambda^i, i=1,\ldots n$, are differentiable functions on (a,b) then λ is a differentiable curve which passes through p. A linear map $\overline{\dot\lambda}_p : F^*(k,p)\to R$ can then be defined via

$$\overline{\dot\lambda}_p(f) \equiv \frac{d}{dt} f(\lambda(t))\Big|_{t=c} = \frac{\partial(f\circ\phi^{-1})}{\partial x^i}\Big|_p \frac{d\lambda^i}{dt}\Big|_c$$

so that $\overline{\dot\lambda}_p(f) = 0$ if $f\in Z^*(k,p)$. Factoring the linear map $\overline{\dot\lambda}_p$ through d_p we get $\overline{\dot\lambda}_p = \dot\lambda_p \circ d_p$ where, clearly, $\dot\lambda_p$ is a linear map from M_p^* into R so that $\dot\lambda_p$ (the <u>tangent vector</u> of λ at p) is an element of M_p.

As $\langle \dot{\lambda}_p, d_p f \rangle = \left.\frac{\partial(f\circ\phi^{-1})}{\partial x^i}\right|_p \left.\frac{d\lambda^i}{dt}\right|_c$ the components of $\dot{\lambda}_p$ relative to the basis $\{\left.\frac{\partial}{\partial x^i}\right|_p, i = 1,\ldots,n\}$ are $\left.\frac{d\lambda^i}{dt}\right|_c$, $i = 1,\ldots,n$. If $\underset{\sim}{v} = v^i \left.\frac{\partial}{\partial x^i}\right|_p$ is in M_p then the curve λ: $(a,b) \to M$ whose representation functions via ϕ are

$$\lambda^i(t) = \phi^i(p) + (t-c)v^i;\ i = 1,2,\ldots,n$$

passes through p and satisfies $\dot{\lambda}_p = \underset{\sim}{v}$. In this way we may characterize the natural basis $\{\left.\frac{\partial}{\partial x^i}\right|_p, i = 1,\ldots,n\}$ of M_p as the set of tangent vectors to the <u>coordinate</u> <u>curves</u> λ_i $(i=1,\ldots,n)$ which pass through p; they are defined relative to the local coordinate system ϕ via the representations

$$\lambda_i^j(t) = \phi^j(p) + (t-c)\delta_i^j;\ i,j = 1,2,\ldots n$$

If we define the tensor space $T^{r,s}(p)$ at $p\in M$ by $T^{r,s}(p) = \underbrace{M_p \otimes \ldots \otimes M_p}_{r} \otimes \underbrace{M_p^* \otimes \ldots \otimes M_p^*}_{s}$ and let (U,ϕ) be a chart of p then the product basis

$$\{\left.\frac{\partial}{\partial x^{i_1}}\right|_p \otimes\ldots\otimes \left.\frac{\partial}{\partial x^{i_r}}\right|_p \otimes d_p x^{j_1} \otimes\ldots\otimes d_p x^{j_s},\ i,j = 1,2,\ldots n\}$$

is the natural basis of ϕ for $T^{r,s}(p)$; any tensor $t\in T^{r,s}(p)$ has, therefore, the component from $t = t_{j_1\ldots j_s}^{i_1\ldots i_r} \frac{\partial}{\partial x^{i_1}}\otimes\ldots\otimes d_p x^{j_s}$.

Further information concerning the tensor algebra associated with the tangent space M_p will be injected as we require it; we assume that the reader is already familiar with the concepts of tensor product, exterior product, etc.

While we shall not require much information concerning the properties of differentiable maps between two differentiable manifolds we do need to know how to define the gradient of such a map. If (M,F_1), (N,F_2) are two differentiable manifolds of class k and dimensions m and n, respectively, then a continuous map $\psi: M \to N$ is said to be differentiable if for every $f_2 \in F_2$, $f_2 \circ \psi \in F_1$ (or $F_2^*(k,\psi(p)) \circ \psi \subset F_1^*(k,p)$ for all $p \in M$). This later relation induces a linear map $\overline{\psi}_p^*: F_2^*(k,\psi(p)) \to F_1^*(k,p)$ if we define $\overline{\psi}_p^*(f_2) = f_2 \circ \psi$ for all $f_2 \in F_2^*(k,\psi(p))$. It is easy to verify that

$$\overline{\psi}_p^*(Z_N^*(k,\psi(p))) \subset Z_M^*(k,p)$$

so that there exists an induced linear map $\psi_p^*: N_{\psi(p)}^* \to M_p^*$ such that $d_p(\overline{\psi}_p^*(f_2)) = \psi_p^*(d_{\psi(p)} f_2)$. The transpose of the linear map ψ_p^* is the linear map $\psi_{*p}: M_p \to N_{\psi(p)}$, called the <u>gradient</u> of ψ at p, which is defined by $\langle \psi_{*p}\underset{\sim}{v}, \underset{\sim}{w}^* \rangle = \langle \underset{\sim}{v}, \psi_p^* \underset{\sim}{w}^* \rangle$ for all $\underset{\sim}{v} \in M_p$ and all $\underset{\sim}{w}^* \in N_{\psi(p)}^*$. For representations of ψ_{*p}, ψ_p^* in terms of local coordinate systems defined near p and $\psi(p)$ we refer the reader to Wang [22].

3. Fibre Bundles, Associated Principal Bundles, and Some Examples.

We recall, first of all, the following

Definition I-2. A Lie group is a group G which has the structure of a C^∞ manifold and is such that the mapping of $G\times G \to G$ defined by $(x,y) \to xy^{-1}$, $x,y\in G$, is C^∞.

and we also need

Definition I-3. If M is a differentiable manifold and G is a Lie group then a differentiable map $L\colon G\times M \to M$ defines G as a Lie transformation group on M if L satisfies

(i) $L(e,m) = m$

(ii) $L(g, L(\tilde{g},m)) = L(g\tilde{g},m)$

for all $g,\tilde{g}\in G$, $m\in M$, where e is the identity element of G.

Following Wang [22] we write $L(g,m) = L_g m = gm$ and call L the operation of left-multiplication of G on M. Clearly $(Lg)^{-1} = Lg^{-1}$ so that L_g is a diffeomorphism of M for each $g\in G$. Left-multiplication of G on M is said to be effective if "L_g acts as the identity map on M" implies that $g = e$. As a simple example we recall that the matrix product defines the general linear group $GL(n,R)$ as a Lie transformation group on $M = R^n$. Also, each Lie group can be considered as a Lie transformation group acting on itself via left-multiplication. We have now collected the basic facts that we need in order to state

Definition I-4. (Wang [22], Chp VI) A fibre bundle L is a collection consisting of three differentiable manifolds, the bundle space L, the base space M, and the fibre space N, a Lie group G called the structure group which is a Lie transformation group acting effectively on N, a smooth map $\pi: L \to M$ called the projection, and a collection of charts $\Phi = \{(U_\alpha, \phi_\alpha), \alpha \in I\}$ called the bundle atlas which satisfies

(i) the elements (U_α, ϕ_α), called bundle charts, consist of open sets $U_\alpha \subset M$ and diffeomorphisms $\phi_\alpha: U_\alpha \times N \to \pi^{-1}(U_\alpha)$ such that $\phi_\alpha(\{p\} \times N) = \pi^{-1}(p)$ for all $p \in U_\alpha$. Each ϕ_α then defines a field of diffeomorphisms on U_α via $\phi_{\alpha,p}: N \to \pi^{-1}(p) \equiv L_p$ and we call L_p the fibre at p.

(ii) On the overlaps $U_\alpha \cap U_\beta$ the mappings $g_{\alpha\beta}(p) \equiv \phi_{\alpha,p}^{-1} \circ \phi_{\beta,p}: N \to N$ belong to G and, furthermore, the fields $g_{\alpha\beta}: U_\alpha \cap U_\beta \to G$ are smooth; these fields are the coordinate transformations on $U_\alpha \cap U_\beta$.

(iii) The set Φ is not a proper subset of any other collection of charts which satisfies (i) and (ii) above, i.e., Φ is maximal relative to (i) and (ii).

Finally, we state the following

Definition I-5. A fibre bundle whose structure group coincides with the fibre space is called a principal bundle.

It can be shown that if L is an arbitrary fibre bundle there exists a principal bundle say P, whose base space, structure group, and coordinate transformations are identical to those of L. (for a proof we refer the reader to Wang [22], Chp. VI); such a bundle is called the associated principal bundle of L.

Examples

I. The Tangent Bundle $T(M)$

The base space is M and the bundle space $T(M) = \bigcup_{p\in M} M_p$. Thus the projection π is a map $\pi\colon T(M) \to M$ such that for $p\in M$, $\pi(p,\underset{\sim}{v}) = p$ where $\underset{\sim}{v}\in M_p$. If we set $T(U) = \pi^{-1}(U) = \bigcup_{p\in U} M_p$ then $M_p = \pi^{-1}(p)$ is called the fibre at p.

The bundle atlas $\{(U_\alpha,\phi_\alpha),\ \alpha\in I\}$ consists of bundle charts (U_α,ϕ_α) such that $\phi_\alpha\colon U_\alpha x R^3 \to T(U_\alpha)$. Hence $\phi_\alpha\colon \{p\}xR^3 \to \pi^{-1}(p)$ and there exist maps $\phi_{\alpha,p}\colon R^3 \to M_p (=\pi^{-1}(p))$; R^3 is called the fibre space. On $U_\alpha \cap U_\beta$, $G_{\alpha\beta}(p) \equiv \phi^{-1}_{\alpha,p}\circ\phi_{\alpha,\beta}\colon R^3 \to R^3$, $p\in U_\alpha\cap U_\beta$, are called the coordinate transformations and we require that, $\forall p\in M$, $G_{\alpha\beta}(p)\in GL(3)$. Thus the general linear group, GL(3), is the structure group of $T(M)$.

Let $\{(U_\alpha,\psi_\alpha),\ \alpha\in I\}$ be an atlas for M such that $\psi_\alpha\colon M \to R^3$ induces local coordinates (x^i) on M, i.e. $\psi_\alpha(p) = (x^1(p),x^2(p),x^3(p))$. We define the ϕ_α above by: $\phi_\alpha(p,v^1,v^2,v^3) = (p,\underset{\sim}{v})$ where $\underset{\sim}{v} = v^i \left.\frac{\partial}{\partial x^i}\right|_p$.

If $\{(U_\alpha,\phi_\alpha),\ \alpha\in I\}$ is maximized w.r.t. all atlases $\{(U_\alpha,\psi_\alpha),\ \alpha\in I\}$ for M, we get the bundle atlas for $T(M)$, say, Φ. Then elements $(U_\alpha,\phi_\alpha)\in\Phi$ give rise to local coordinate systems $(x^1,x^2,x^3,v^1,v^2,v^3)$ on $T(U_\alpha)$, called the lifted coordinate systems. If ψ_β induces local coordinates $(\bar{x}^i)$

on U_β and $p \in U_\alpha \cap U_\beta$ then the coordinate transformations are given by $G_{\alpha\beta}(p) = \det[\partial x^i / \partial x^{-j}]$.

II. The Bundle of Linear Frames $E(M)$

$E(M)$ is the associated principal bundle of $T(M)$. A linear frame at p is an ordered basis for M_p, i.e., $e_p = \{e_{p,i};\ i=1,2,3\}$. Set $E_p = \{e_p\}$ then the base space is again M and the bundle space is $E(M) = \bigcup_{p \in M} E_p$; the projection $\pi: E(M) \to M$ such that $\pi(p,e_p) = p$, $\forall p \in M$. If we set $E(U) = {}_{p \in U} E_p = \pi^{-1}(U)$ then $E_p = \pi^{-1}(p)$ is the fibre at p.

The bundle atlas $\xi = \{(U_\alpha, \xi_\alpha),\ \alpha \in I\}$ consists of maps $\xi_\alpha: U_\alpha \times GL(3) \to E(U_\alpha)$. Thus, as with $T(M)$, there exist maps $\xi_{\alpha,p}: GL(3) \to E_p(=\pi^{-1}(p))$. If $\underset{\sim}{G} \in GL(3)$ has the representation $\underset{\sim}{G} = [G^i_j]$ then the operation of <u>right multiplication by $\underset{\sim}{G}$ on $E(M)$</u> is defined by $R_{\underset{\sim}{G}} e_p = \{e_{p,j} G^j_i;\ i=1,2,3\}$.

Let $\phi = \{(U_\alpha, \phi_\alpha),\ \alpha \in I\}$ be a bundle atlas for $T(M)$ and $i = \{(1,0,0), (0,1,0), (0,0,1)\}$ the standard basis for R^3. Define $e_p(\alpha) = \phi_{\alpha,p}(i)$, then $e_p(\alpha)$ is a frame at p since $\phi_{\alpha,p}: R^3 \to M_p$ is an isomorphism. We now define the map ξ_α by $\xi_\alpha(p,\underset{\sim}{G}) = R_{\underset{\sim}{G}}(e_p(\alpha))$ and note that it is easy to show that the coordinate transformations on $T(M)$ and $E(M)$ coincide.

4. Lie Algebras, the Exponential Map, and Fundamental Fields on $E(M)$.

Every Lie group G has an associated Lie algebra g, i.e., a vector space which is equipped with a bracket operation. To define this Lie algebra, let $\underset{\sim}{v}$ be a vector field on G, and let $L_x y = xy$, $\forall x,y\epsilon G$ (i.e. we consider G as a Lie transformation group acting on itself via left-multiplication so that $L_x: G \to G, \forall x\epsilon G$; if $L_{x*y}(\underset{\sim}{v}(y)) = \underset{\sim}{v}(xy)$ for all $x,y\epsilon G$ then $\underset{\sim}{v}$ is said to be a left-invariant vector field and the collection g of all such left-invariant vector fields on G then forms a vector space in the obvious way. To define the bracket operation on g we first define the Lie derivative of one vector field $\underset{\sim}{u}$ with respect to another $\underset{\sim}{v}$ as follows: if M is a differentiable manifold, $p\epsilon M$, and ϕ a local coordinate system near p so that $\underset{\sim}{u} = u^i \frac{\partial}{\partial x^i}\Big|_p$, $\underset{\sim}{v} = v^i \frac{\partial}{\partial x^i}\Big|_p$ then

$$[L_{\underset{\sim}{v}} \underset{\sim}{u}](p) = \left(\frac{\partial u^i}{\partial x^j}(p)v^j(p) - \frac{\partial v^i}{\partial x^j}(p)u^j(p)\right)\frac{\partial}{\partial x^i}\Big|_p$$

The Lie derivative of any arbitrary tensor $\underset{\sim}{y}\epsilon T^{r,s}(p)$ with respect to $\underset{\sim}{v}$ can be defined in a coordinate-free manner if we regard $\underset{\sim}{v}$ as the infinitesimal generator of a one-parameter family of differentiable maps ψ of a neighborhood of p (i.e., see Wang [22], Chp. III) but the above definition for vector fields shall suffice for our purposes. If $\underset{\sim}{u},\underset{\sim}{v}\epsilon g$ then we define $[\underset{\sim}{v},\underset{\sim}{u}] = L_{\underset{\sim}{v}}\underset{\sim}{u}$ and this bracket operation endows g with the structure of a Lie algebra. It can also be shown that there exists an isomorphism $|_e : \underset{\sim}{g} \to G_e$ such that $|_e : \underset{\sim}{v} = \underset{\sim}{v}(e)$ if $\underset{\sim}{v} \in \underset{\sim}{g}$.

If G is a Lie group and $\underset{\sim}{v} \in \underset{\sim}{g}$ then $\underset{\sim}{v}$ induces a one-parameter subgroup $\lambda_{\underset{\sim}{v}}(t)$ of G such that $\partial_t \lambda_{\underset{\sim}{v}}(t)|_{t=0} = \underset{\sim}{v}(e)$. In fact we may take $\lambda_{\underset{\sim}{v}}(t)$ as the solution of $\dot{\lambda}(t) = \underset{\sim}{v}(\lambda(t))$ satisfying $\lambda(0) = e$ so that $\dot{\lambda}(0) = v(e)$, i.e., $\lambda_{\underset{\sim}{v}}$ is an integral curve of the vector field $\underset{\sim}{v}$. The <u>exponential map</u>, exp: $\underset{\sim}{g} \to G$, is then defined by $\exp(\underset{\sim}{v}) = \lambda_{\underset{\sim}{v}}(1)$ for all $\underset{\sim}{v} \in \underset{\sim}{g}$. For $G \equiv GL(3)$, $\lambda_{\underset{\sim}{V}}(t) = 1 + \underset{\sim}{V}t + \frac{1}{2}\underset{\sim}{V}t^2 + \ldots$, where $\underset{\sim}{V} \in g\ell(3)$, so that $\exp V = 1 + \underset{\sim}{V} + \frac{1}{2}\underset{\sim}{V}^2 + \ldots$. For $\underset{\sim}{V} \in g\ell(3)$ and $G \in GL(3)$ it is possible to prove that $\exp(G\underset{\sim}{V}G^{-1}) = G\exp(\underset{\sim}{V})G^{-1}$ and, also, that $\det[\exp(\underset{\sim}{V})] = \exp(\mathrm{tr}\underset{\sim}{V})$ where $\mathrm{tr}\underset{\sim}{V}$ denotes the trace of $\underset{\sim}{V}$.

Finally, let G denote the structure group for $E(M)$(1) and $\xi = \{(U_\alpha, \xi_\alpha), \alpha \in I\}$ the bundle atlas. Then $\xi_{\alpha,p}: G \to E_p$ and hence there exist maps $\xi_{\alpha,p*}: G_x \to (E_p)_x$, i.e., if $\underset{\sim}{v} \in \underset{\sim}{g}$ then $\xi_{\alpha,p*}(\underset{\sim}{v}) = \bar{v}_\alpha(p)$ which is a vector field on E_p. It can be shown that these fields are independent of the bundle chart; $\underset{\sim}{\bar{v}}$ is called a <u>fundamental</u> <u>field</u> on $E(M)$ and we denote the set of all such $\underset{\sim}{\bar{v}}$ by $\bar{g}$. Now let $\pi: E(M) \to M$ so that $\pi_*: E(M)_x \to M_{\pi(x)}$. Then it is possible to show that $\pi_*(\underset{\sim}{\bar{v}}) = 0$, $\forall \underset{\sim}{\bar{v}} \in \underset{\sim}{\bar{g}}$ and we say that the $\underset{\sim}{\bar{v}}$ lie in the fibre directions; in fact it is possible to show that $\underset{\sim}{\bar{g}}$ is isomorphic to ker π_*.

(1) in this case, of course, $G = GL(3)$ and the associated Lie algebra $\underset{\sim}{g} = gl(3)$.

5. "G" Connections on $E(M)$ and Parallel Transport

Let $x \in E(M)$. Then the vertical subspace V_x of the tangent space $E(M)_x$ is defined to be $V_x = \ker \pi_{*x}$, i.e., $V_x = \bar{g}|_x$. A connection on $E(M)$ is then a map $H: x \to H_x \subset E(M)_x$ such that $E(M)_x = V_x \oplus H_x$, $x \in E(M)$. We note that the horizontal subspace at x, i.e. H_x, is not unique. Since $V_x = \ker \pi_{*x}$ we have $\pi_{*x}|_{H_x} \equiv \eta_x: H_x \to M_{\pi(x)}$ is an isomorphism. Thus $\forall \underset{\sim}{v} \in M_{\pi(x)}$, there exists $\bar{\underset{\sim}{v}}$ on E_p such that $\bar{\underset{\sim}{v}}(x) = \eta_x^{-1}(\underset{\sim}{v})$. The point $\bar{\underset{\sim}{v}}$ is called the horizontal lift of $\underset{\sim}{v}$ relative to H.

Let λ be a smooth curve in M. The horizontal lift of λ is a curve $\tilde{\lambda} \in E(M)$ such that $\pi(\tilde{\lambda}(t)) = \lambda(t)$, $\forall t \in [a,b]$, say, and such that $\tilde{\lambda}$ is horizontal, i.e., the tangent vectors $\partial_t \tilde{\lambda}(t)$ are the horizontal lifts of the tangent vectors $\partial_t \lambda(t)$, $\forall t \in [a,b]$. Thus there exist maps $\rho_t(\tilde{\lambda}): E(M)_{\lambda(0)} \to E(M)_{\lambda(t)}$ called the parallel transports along λ relative to the connection H.

Now, let (U_α, ξ_α) be a lifted bundle chart for $E(M)$ such that $\lambda \in U_\alpha$. Define $\rho_{t,\alpha}: G \to G$ by $\rho_{t,\alpha} \equiv \xi^{-1}_{\alpha,\lambda(t)} \circ \rho_t(\tilde{\lambda}) \circ \xi_{\alpha\ \lambda(0)}$. Then, H is called a "G" connection on $E(M)$ if, for all smooth $\lambda \in M$, $\rho_{t,\alpha} \in G$[(2)]. It can be shown that H is a "G" connection iff $\rho_t(\tilde{\lambda})$ is independent of $\tilde{\lambda}$ so that there exist well defined parallel transports of the tangent spaces along λ, i.e. since a linear frame, at $p \in M$, is an ordered basis for M_p, we may extend ρ_t to $\hat{\rho}_t: M_{\lambda(0)} \to M_{\lambda(t)}$, which is a linear isomorphism. Thus "G" connections on $E(M)$ correspond

(2) thus, a "G" connection H on $E(M)$ is a connection for which the maps $\rho_{t,\alpha} \in GL(3)$; "G" connections on arbitrary principal bundles are defined in an analogous way.

to the <u>classical</u> <u>affine</u> <u>connections</u> on $T(M)$.

Once again, let (U_α, ξ_α) be a <u>lifted</u> <u>chart</u> for $E(M)$ corresponding to the local coordinate system (x^1, x^2, x^3) on M. Then $(p, e_p) \in M(U_\alpha)$ has local coordinates $(x^i(p),\ e^j_k(p))$ relative to (U_α, ξ_α), where $e_p = \{e^j_k(p) \frac{\partial}{\partial x^j}\Big|_p\}$. Put $x = (p, e_p)$. Then $E(M)_x$ is spanned by the natural basis: $\{\frac{\partial}{\partial x^i}\Big|_x, \frac{\partial}{\partial e^j_k}\Big|_x\}$.

If we compute the matrix π_* and apply it to this set we find that V_x is spanned by $\{\frac{\partial}{\partial e^j_k}\Big|_x\}$ and thus H_x is spanned by $\{\frac{\partial}{\partial x^i}\Big|_x - \Gamma^j_{ki}(p, e_p)\frac{\partial}{\partial e^j_k}\Big|_x\}$, where the Γ^j_{ki} are the connection symbols of H. Now, let $\tilde{\lambda}(t) = (\lambda(t), e^j_k(t)\partial/\partial x^j|\lambda(t))$ be the representation of the horizontal lift of $\lambda(t) \in M$. Relative to the lifted chart (U_α, ξ_α), $\tilde{\lambda}(t) = (\lambda^i(t), e^j_k(t))$, and thus $\partial_t\tilde{\lambda}(t) = (\partial_t\lambda^i(t), \partial_t e^j_k(t))$. If we set the vertical components of $\partial_t\tilde{\lambda}(t) = 0$ then we find the <u>equations</u> <u>of</u> <u>parallel</u> <u>transport</u>, namely, $\partial_t e^i_j(t) + \Gamma^i_{kj}\partial_t\lambda^k = 0$. The solution of this equation, with appropriate initial conditions, is such as to render $\tilde{\lambda}(t) = (\lambda(t), e^j_k(t)\partial/\partial x^j|\lambda(t))$ a <u>horizontal</u> <u>curve</u>. Finally we can show, by virtue of the fact that we are dealing with a "G" connection that $\Gamma^j_{ki}(p, e_p) = \Gamma^j_{ki}(p)e^\ell_k$, $\forall p \in M$.

6. Convariant Derivatives, Curvature, Torsion, and Flatness

It can be shown that the parallel transports, $\hat{\rho}_t\colon M_{\lambda(0)} \to M_{\lambda(t)}$ induce linear isomorphisms of the tensor spaces: $\hat{\rho}_t^{r,s}\colon T^{r,s}_{\lambda(0)} \to T^{r,s}_{\lambda(t)}$, i.e., $\underset{\sim}{y}(t) = \hat{\rho}_t^{r,s}(\underset{\sim}{y}(0))$ where $\underset{\sim}{y}(t)\varepsilon T^{r,s}_{\lambda(t)}$ and $y(0)\varepsilon T^{r,s}_{\lambda(0)}$. Here, of course,
$T^{r,s}_{\lambda(0)} = \underbrace{M_{\lambda(0)} \otimes \ldots \otimes M_{\lambda(0)}}_{r} \otimes \underbrace{M^*_{\lambda(0)} \ldots \otimes M^*_{\lambda(0)}}_{s}$, where $M^*_{\lambda(0)}$ denotes the dual space of $M_{\lambda(0)}$. Now let $\underset{\sim}{z}(t)\varepsilon T^{r,s}_{\lambda(t)}$; the parallel transports of the tensor spaces then induce the operation of covariant differention,
$\frac{Dz}{dt} = \lim_{\Delta t\to o} 1/\Delta t\{\underset{\sim}{z}(t)-\hat{\rho}_t^{r,s}\underset{\sim}{z}(t-\Delta t)\}$, which says that $\underset{\sim}{z}(t)$ is obtained via parallel transport iff $\frac{Dz}{dt} = 0$. It is then possible to show that if $\tilde{\underset{\sim}{z}}$ is a smooth tensor field defined on some open set in M, $\lambda(t)\ \varepsilon$ Dom $(\tilde{z})$, and $\hat{\underset{\sim}{z}}(t) = \tilde{\underset{\sim}{z}}(\lambda(t))$ is a smooth field, then there exists a tensor field $D\underset{\sim}{z}$ such that $\frac{Dz}{dt} = \langle D\underset{\sim}{z},\dot{\underset{\sim}{\lambda}}\rangle \equiv D_{\dot{\underset{\sim}{\lambda}}}\underset{\sim}{z}$. For instance, if $\underset{\sim}{z}$ is actually a vector field $\underset{\sim}{v}$ then, in terms of local coordinates,

$$\frac{D\underset{\sim}{v}}{dt}\Big|_{\lambda(t)} = [v^i{}_{,k}\big|_{\lambda(t)} + \Gamma^i_{jk}(\lambda(t))v^j(t)]\dot{\lambda}^k(t)\frac{\partial}{\partial x^i}\Big|_{\lambda(t)}$$

$$= ([v^i{}_{,k} + \Gamma^i_{jk}v^j]\frac{\partial}{\partial x^i}\Big|_p \otimes d_p x^k) \circ (\dot{\lambda}^k\frac{\partial}{\partial x^k}\Big|_p)$$

$$= (v^i{}_{,k}\frac{\partial}{\partial x^i}\Big|_p \otimes d_p x^k) \circ (\dot{\lambda}^k\frac{\partial}{\partial x^k}\Big|_p)$$

$$= D\underset{\sim}{v}\big|_{\lambda(t)} \circ \dot{\lambda}\big|_p.$$

Finally we have the following definitions which will be utilized when we come to characterize the material connections which arise in the study of continuous distributions of dislocations in a continua: first of all, an affine connection on $T(M)$ is said to be (locally) flat if if for all $p \varepsilon M$ there exists a local coordinate system (x^i) such that relative to these local coordinates the Γ symbols of the connection vanish. An affine connection is completely integrable if it admits an integral manifold in the neighborhood of p for each $p \varepsilon M$; to make this concept precise recall that there exists a one-to-one correspondence between affine connections on $T(M)$ and G connections on $E(M)$ so that a connection H on $T(M)$ is a smooth field whose values are horizontal subspaces. If H is a connection on $T(M)$ and $x \varepsilon T(M)$ then the map $\eta_x \equiv \pi_{*x}|_{H_x} : H_x \to M_p$ is an isomorphism where $\pi(x) = p$. Now, given any field P of r-dimensional subspaces $(r \leq n)$ of the tangent spaces on M let $i: N \to M$ be an immersion which defines $i(N)$ as a submanifold of M. Then $i(N)$ is called an integral manifold of p if for every point $q \varepsilon N$, $i_{*q}(N_q) \subset P(i(q))$ (if $i(q) = p \varepsilon M$ then $P(p) \subset M_p$ and has dimension $r \leq n$).

The condition of integrability for a G connection H on $E(M)$(equivalently, affine connections on $T(M)$) is characterized by the curvature tensor $\underset{\sim}{R}$ whose components relative to a local coordinate system (x^i) are

$$R^j_{mri} = \frac{\partial \Gamma^j_{mi}}{\partial x^r} - \frac{\partial \Gamma^j_{mr}}{\partial x^i} + \Gamma^i_{sr} \Gamma^s_{mi} - \Gamma^j_{si} \Gamma^s_{mr}$$

where the Γ's are the connection symbols of H. By the famous theorem of Frobenius, a connection H is completely integrable iff $\underset{\sim}{R} = \underset{\sim}{0}$. Now let H be a completely integrable connection on $E(M)$ and let $\mu(p) = \{(p, f_i(p));\ i = 1,2,3\}$ be a horizontal cross-section over $U_\alpha \subset M$ such that $f_i(p) = f_i^j \frac{\partial}{\partial x^i}\Big|_p$ (a cross-section over $U \subset M$ is a map $\sigma: U \to E(U)$ such that $\pi \circ \sigma = id_M$ the identity map on M); in saying that the cross-section is horizontal we imply that the f_j^i satisfy the equations of parallel transport, i.e., $\frac{\partial f_k^i}{\partial x^m} + \Gamma_{im}^j f_k^i = 0$. It then follows that the Poisson brackets $[f_i, f_j]^k = \frac{\partial f_j^k}{\partial x^m} f_i^m - \frac{\partial f_i^k}{\partial x^m} f_j = T_{mn}^k f_i^m f_j^n$ where $T_{mn}^k \equiv \Gamma_{mn}^k - \Gamma_{nm}^k$ are the components of the torsion tensor $\underset{\sim}{T}$ associated with H. If $\underset{\sim}{T} = \underset{\sim}{0}$ then $[f_i, f_j]^k = 0$ and the Frobenius theorem can be used again to infer that there exists a local coordinate system y^i on M such that $f_i = \frac{\partial}{\partial y^i}$ which, in turn, implies that $\Gamma_{jk}^i(y) = 0$, i.e., that H is flat.

Chapter II. Material Uniformity in Elasticity

1. Introduction

We present, in this chapter, a concrete mathematical formulation of the theory of continuous static distributions of dislocations in a simple elastic body; an extension of this theory to cover certain classes of non-simple materials will be given in the following chapter. The problem of determining the material geometric structure of a simple body from the constitutive equations of the particles comprising the body was first treated by Noll; an account of his early work in this direction can be found in the treatise by Truesdell and Noll [5] and a more elaborate treatment, but one in which the material geometric structure is still required to be derivable from a globally smooth distant parallelism, was given by Noll in [6]. The theory, as we shall present it here, is based on a truly remarkable paper of C. C. Wang [7]. This work of Wang's, which appeared at the same time as [6], has the advantage of removing the unnecessarily restrictive smoothness assumption of Noll, which we mentioned above, and thus allows for the treatment of a much wider class of simple bodies; it is probably also the first work of its kind to demonstrate effectively the power of modern differential geometric tools in continuum mechanics.

While we shall continue beyond the static dislocation theory to be presented here, to treat non-simple materials and anelastic response (dislocation motion), some other accounts of the material in this chapter may be found in [24] and [25]. In addition, reprints of the foundation papers by Noll and Wang together with several other papers which treat classes of universal solutions for materially uniform elastic bodies, as well as wave propagation in such materials, are to be found in the collection [26]. Truesdell has included a lucid summary of Noll's basic ideas in this area in his Lectures on Natural Philosophy [27] and other interpretations of the concepts introduced by Noll, as well as comparisons with their own work in dislocation theory, have been given by Bilby [9] and Kröner [10]; in the same collection which contains these latter works the reader may find two brief expositions by Noll and Wang, respectively, of the basic ideas which underlie the theory to be presented below. As far as possible we shall retain the notation of the original papers.

2. Body Manifolds, Motions and Deformation Gradients

We begin with the following,

Definition II-1 (Wang) A body manifold $\mathcal{B}$ is an oriented three-dimensional differentiable manifold which is connected and has the property that there exist diffeomorphisms, say ϕ, ψ, χ,... (which we shall call configurations of the body $\mathcal{B}$) which map $\mathcal{B}$ into R^3, i.e., $\phi\colon \mathcal{B} \to R^3$.

If $p \in \mathcal{B}$, a body manifold, then a linear isomorphism

$\underset{\sim}{r}_p: \mathcal{B}_p \to \mathcal{R}^3$ is called a <u>local</u> <u>configuration</u> of p; of course, as the tangent space $\mathcal{B}_p$ is an oriented three-dimensional vector space it must be algebracially isomorphic to $\mathcal{R}^3$; we note moreover that both configurations and local configurations are required to be orientation preserving and that the gradient of a configuration ϕ of $\mathcal{B}$ gives rises to a field of local configurations of points $p\varepsilon\mathcal{B}$, i.e., if $\phi: \mathcal{B} \to \mathcal{R}^3$ is a configuration of $\mathcal{B}$ then $\phi_{*p}: \mathcal{B}_p \to \mathcal{R}^3$ is a local configuration of p for each $p\varepsilon\mathcal{B}$. The converse is not true as, in general, a given field of local configurations can not be obtained as the gradient of a configuration of $\mathcal{B}$. We now define a <u>motion</u> of $\mathcal{B}$ to be a one-parameter family $\phi(t)$ of configurations of $\mathcal{B}$, where t is a time variable, and a <u>local</u> <u>motion</u> of $p\varepsilon\mathcal{B}$ to be a one-parameter family $\underset{\sim}{r}_p(t)$ of local configurations of p. If we choose to pick out a particular local configuration $\underset{\sim}{r}_p$ of $p\varepsilon\mathcal{B}$ then we will call $\underset{\sim}{r}_p$ a <u>local</u> <u>reference</u> <u>configuration</u>; if $\phi(t)$ is a motion of $\mathcal{B}$ and $\underset{\sim}{r}_p$ is such a local reference configuration of p then we can define the tensor

$$\underset{\sim}{F}_p(t) \equiv \phi_{*p}(t) \circ \underset{\sim}{r}_p^{-1}, \quad t > 0$$

which we term the <u>local</u> <u>deformation</u> at p (relative to $\underset{\sim}{r}_p$). Now, even though we may not be able to find configurations ψ of $\mathcal{B}$ such that $\underset{\sim}{r}_p = \psi_{*p}$, $\forall p\varepsilon\mathcal{B}$, if we fix $\underset{\sim}{p}\varepsilon\mathcal{B}$ then certainly such a relationship can be satisfied at this one point. The chain rule for gradients then yields

$$\underset{\sim}{F}_p(t) = \phi_{*p}(t) \circ \psi_{*p}^{-1} = [\phi(t) \circ \psi^{-1}]_{*\psi(p)}$$

and as $\phi(t) \circ \psi^{-1}$: $\psi(\mathcal{B}) \to [\phi(t)](\mathcal{B})$ represents a deformation of the (open) domain $\psi(\mathcal{B})$ in $\mathcal{R}^3$ we also call $\underset{\sim}{F}_p(t)$ the deformation gradient at the point at time t. As F is an orientation preserving isomorphism of $\mathcal{R}^3$ we have det $F(t)>0$ for all $t>0$.

If ϕ: $\mathcal{B} \to \mathcal{R}^3$ is a configuration of $\mathcal{B}$ then ϕ can be characterized by three smooth functions $x^i(p)$, $p\varepsilon\mathcal{B}$, i = 1,2,3, such that

$$\phi(p) = (x^1(p),\ x^2(p),\ x^3(p)), \qquad p\varepsilon\mathcal{B}$$

where the x^i are, of course, the coordinate functions of ϕ. Now let $\{\underset{\sim}{i}_a,\ a = 1,2,3\}$ denote the vectors {(1,0,0), (0,1,0), (0,0,1)} which comprise the standard basis of $\mathcal{R}^3$. If $\underset{\sim}{r}_p$: $\mathcal{B}_p \to \mathcal{R}^3$ is a local configuration of $p\varepsilon\mathcal{B}$ it can be characterized by a basis $\{\underset{\sim}{e}_a,\ a = 1,2,3\}$ in $\mathcal{B}_p$ such that $\underset{\sim}{r}_p(\underset{\sim}{e}_a) = \underset{\sim}{i}_a$, a = 1,2,3. We call $\{\underset{\sim}{e}_a,\ a = 1,2,3\}$ the reference basis of $\underset{\sim}{r}_p$ and note that if ϕ is a configuration of $\mathcal{B}$ then the reference basis of ϕ_{*p} is just the natural basis $\{\frac{\partial}{\partial x^i}\big|_p,\ i = 1,2,3\}$.

Now let κ: $\mathcal{B} \to \mathcal{R}^3$ be a particular configuration of $\mathcal{B}$ which we shall single out and use as a reference configuration; we denote the coordinate functions of κ by X^A, A = 1,2,3. If ϕ is any other configuration of $\mathcal{B}$ then the deformation from κ to ϕ is a diffeomorphism

$$\phi \circ \kappa^{-1}: \kappa(\mathcal{B}) \to \phi(\mathcal{B})$$

which, as $\kappa(\mathcal{B})$ and $\phi(\mathcal{B})$ are open connected sets in $\mathcal{R}^3$, can be characterized by the deformation functions $x^i = x^i(X^A)$, $i = 1,2,3$, in such a way that $\phi \circ \kappa^{-1}(X^A) = x^i$, $\forall X^A \varepsilon \kappa(\mathcal{B})$. The deformation functions are smooth since both ϕ and κ are smooth maps. If ϕ is a motion of $\mathcal{B}$ then the above representation becomes $x^i = x^i(X^A,t)$, $i = 1,2,3$.

Now let κ be a reference configuration of $\mathcal{B}$ and $\phi(t)$ a motion of $\mathcal{B}$. Then κ_{*p} is a local reference configuration of $p\varepsilon\mathcal{B}$ for all such p and the local deformation at p relative to κ_{*p} is $F_p(t) = \phi_{*p}(t) \circ \kappa_{*p}^{-1} = (\phi \circ \kappa^{-1})_{*\kappa(p)}$. We can characterize $\underset{\sim}{F}_p$ in component form via $\underset{\sim}{F}_p i_A = F^b_A i_b$, $A = 1,2,3$. But in terms of the reference bases of κ_{*p} and ϕ_{*p}, which are just the natural bases $\{\frac{\partial}{\partial X^A}, A = 1,2,3\}$ and $\{\frac{\partial}{\partial x^a}, a = 1,2,3\}$, respectively, we have

$$\phi_{*p}^{-1} \circ \underset{\sim}{F}_p \underset{\sim}{i}_A = F^b_A \phi^{-1}_{*p} \underset{\sim}{i}_b = F^b_A \frac{\partial}{\partial x^b} = \kappa^{-1}_{*p} \underset{\sim}{i}_A = \frac{\partial}{\partial X^A}$$

so that $F^b_A = \frac{\partial x^b}{\partial X^A}$. This completes as much of the material dealing with kinematics of a continua as we shall need for now.

3. Force and Stress in Continuum Mechanics

In order to formulate the constitutive equation of an elastic point we must first understand what we mean when we talk about the concept of stress. To this end, let $\mathcal{C}$ be a part of a body manifold $\mathcal{B}$ which is in the configuration κ. We assume here that besides being a three-dimensional differentiable manifold, $\mathcal{B}$ is also a measure space, i.e., it is endowed with a non-negative scalar measure which is called the mass distribution of the body; a part C of $\mathcal{B}$ is then any measurable subset of $\mathcal{B}$ which can be mapped onto a region in R^3. It is also assumed that for any configuration of $\mathcal{B}$, such as κ, the induced measure over $\kappa(\mathcal{B})$ is absolutely continuous and thus (by the Radoń - Nikodym theorem) has a density ρ_κ which is called the mass density of $\mathcal{B}$ in the configuration κ. Thus if $C \subset \mathcal{B}$ is a part in $\mathcal{B}$

$$m(C) = \int_{\kappa(C)} \rho_\kappa dv (\int \equiv \text{Lebesque integral})$$

where v is the Euclidean volume measure on R^3.

Of paramount importance in modern continuum mechanics is the concept of force which describes the action of the outside world on a body in motion and, also, the interaction between the different parts of the body. It is usually assumed that mutual body forces are absent and that all forces are continuously distributed. Following Truesdell and Noll [5] we now list below those conditions which characterize a system of forces for the body $\mathcal{B}$ in any motion $\phi(t)$:

(i) At each time t, a vector field $\underset{\sim}{b}(x,t)$, called the density of the external body force per unit mass acting on $\mathcal{B}$ in the motion ϕ is defined for each $\underset{\sim}{x} \varepsilon \phi_t(\mathcal{B})$; by $\phi_t(\mathcal{B})$ we indicate the configuration occupied by $\mathcal{B}$ at time t in the motion $\phi(t)$. The vector $\underset{\sim}{f}_b(C)$, defined to be the integral $\int_{\phi_t(C)} \underset{\sim}{b}(\underset{\sim}{x},t)dm$ is called the resultant external body force exerted on the part C at time t.

(ii) At any time t, there corresponds to each part $C \subset \mathcal{B}$ a vector field $\underset{\sim}{t}(\underset{\sim}{x}; C)$ acting on C, called the <u>stress</u>, which is defined for all points $\underset{\sim}{x} \varepsilon \phi_t(\partial C)$. The stress is a measure of the density of the <u>contact force</u> and the resultant contact force $\underset{\sim}{f}_c(C)$ exerted on C at time t is given by the surface integral $\oint_{\phi_t(\partial C)} \underset{\sim}{t}(\underset{\sim}{x}; C)dS$

(iii) The <u>total resultant force</u> $f(C)$ acting on $C \subset \mathcal{B}$ is given by: $\underset{\sim}{f}(C) \equiv \underset{\sim}{f}_b(C) + \underset{\sim}{f}_c(C)$.

(iv) There is a vector-valued function $\underset{\sim}{t}(\underset{\sim}{x},\underset{\sim}{n})$ defined for all points $\underset{\sim}{x}$ in $\phi_t(C)$ and all unit vectors $\underset{\sim}{n}$ such that the stress acting on C is given by

$$\underset{\sim}{t}(\underset{\sim}{x}; C) = \underset{\sim}{t}(\underset{\sim}{x},\underset{\sim}{n})$$

where $\underset{\sim}{n}$ is the exterior unit normal at the point $\underset{\sim}{x}$ on the boundary of C in the configuration ϕ_t. We call $\underset{\sim}{t}(\underset{\sim}{x},n)$ the <u>stress vector</u> at $\underset{\sim}{x}$ and the above concept is known as the <u>stress principle</u> of Cauchy.

Any system of forces such as that specified above must obey the fundamental balance laws of mechanics, i.e., the principles of balance of momentum and balance of moment of momentum. If the momentum $M(C)$ of C in the configuration ϕ_t is given by $\int_{\phi_t(C)} \dot{\underset{\sim}{x}}\, dm$ then the principle of balance of momentum requires that $\underset{\sim}{f}(C) = \dot{M}(C)$ and implies, under suitable continuity conditions, the existence of a tensor field $\underset{\sim}{T}_S$ such that $\underset{\sim}{t}(\underset{\sim}{x},\underset{\sim}{n}) = \underset{\sim}{T}_S(\underset{\sim}{x})\underset{\sim}{n}$ for all $x \varepsilon \phi_t(\partial C)$; the superposed dot above indicates, of course, differentiation with respect to time. We call $\underset{\sim}{T}_S(\underset{\sim}{x})$ the stress tensor at the point $\underset{\sim}{x}$ and it is relatively simple to show that balance of momentum then implies the Cauchy law of motion:

$$\text{div } \underset{\sim}{T}_S + \rho \underset{\sim}{b} = \rho \ddot{\underset{\sim}{x}}$$

while balance of moment of momentum implies the symmetry of the stress tensor, i.e., $\underset{\sim}{T}_S = \underset{\sim}{T}^t_S$.

4. The Constitutive Equation of a Simple Elastic Point; Material Isomorphisms and Materially Uniform Elastic Bodies.

By a constitutive equation in continuum mechanics we understand a relation between the contact forces, acting on parts of the body $\mathcal{B}$, which are specified by the stress tensor, and motions of $\mathcal{B}$; constitutive equations thus serve as representations of various classes of ideal materials. All constitutive equations which are dealt with in modern continuum mechanics must satisfy three basic principles, i.e., the principle of determinism, the principle of local action, and the principle of material frame-indifference; the later principle asserts, essentially, that the material properties of a body, i.e., its response to the application of given forces, are indifferent to the frame of reference used by an observer. The first two principles are automatically satisfied by a body which consists of simple elastic particles (points) and while the restrictions on the form of the constitutive function which are implied by the third principle above are both important and interesting we shall have no occasion to make use of them in the present work. For excellent discussions of these principles and their ramifications, particularly within the scope of the theory of simple materials, we refer the reader to either of the presentations to be found in [5] or [27].

Recall now that the contact forces at a point $\underset{\sim}{x} = \phi(p)$, in a configuration ϕ of $\mathcal{B}$, $p \in \mathcal{B}$, are characterized by the stress tensor $\underset{\sim}{T}_S$, which is a symmetric tensor of order two on R^3; this tensor acts on the unit normal at points of $\phi(\partial C)$, where $C \subset \mathcal{B}$ is a <u>part</u> of $\mathcal{B}$, and transforms it into the stress vector at that point. Instead of denoting this tensor by $T_S(\underset{\sim}{x}) \equiv T_S(\phi_t(p))$ as we have in the last section we shall usually just write $\underset{\sim}{T}_S(p,t)$(for the stress tensor at the particle p at time t in the motion ϕ) if the motion $\phi(t)$ is understood.

Our basis definition is the one given by Noll [6]:

<u>Definition</u> II-2. If $p \in \mathcal{B}$ and $\phi(t)$ is any motion of $\mathcal{B}$ then p is called a simple elastic particle if the stress tensor at p at time t in the motion ϕ depends only on $\phi_{*p}(t)$, i.e.,

there exists a tensor function $\tilde{E}$ such that

$$\tilde{T}(p,t) = \tilde{E}(\phi_{*p}(t),\ p)$$

We call the tensor function $\tilde{E}$ the elastic response function. If $\tilde{r}_p$ is a fixed local reference configuration of p then we have $\phi_{*p}(t) = \tilde{F}(t) \circ \tilde{r}_p$ and

$$\tilde{T}(p,t) = \tilde{E}(\tilde{F}(t) \circ \tilde{r}_p, p) \equiv \tilde{S}_{\tilde{r}_p}(\tilde{F}(t),\ p) \tag{II-1}$$

where $\tilde{S}_{\tilde{r}_p}$ is the <u>elastic response function relative</u> to $\tilde{r}_p$; in what follows we shall suppress the dependence of $\tilde{S}$ on $\tilde{r}_p$. Following Noll [6] we now make

<u>Definition</u> II-3. Two simple elastic particles $p,q \varepsilon \mathcal{B}$ are called <u>materially isomorphic</u> if there exist local reference configurations $\tilde{r}_p$ and $\tilde{r}_q$ such that

$$\tilde{S}(\tilde{F},p) \equiv \tilde{S}(\tilde{F},q), \quad \forall \tilde{F} \varepsilon GL(3) \tag{II-2}$$

From the definition it is a simple matter to derive

<u>Theorem</u> II-1. If $p,q \varepsilon \mathcal{B}$ are simple elastic particles then they are materially isomorphic iff there exists an isomorphism $\tilde{r}(p,q)$: $\mathcal{B}_p \to \mathcal{B}_q$ such that

$$\tilde{E}(\phi_{*q},q) \equiv \tilde{E}(\phi_{*q} \circ \tilde{r}(p,q),\ p) \tag{II-3}$$

for all configurations ϕ: $\mathcal{B} \to R^3$.

An isomorphism $\tilde{r}(p,q)$: $\mathcal{B}_p \to \mathcal{B}_q$ with the above property is termed a <u>material isomorphism</u> of p and q. We also have the following result of Noll [6]:

Theorem II-2 If $\underset{\sim}{r}_p$, $\underset{\sim}{r}_q$ are local reference configurations of p,q$\epsilon\mathcal{B}$, respectively, such that

$$\underset{\sim}{r}(p,q) \equiv \underset{\sim}{r}_q^{-1} \circ \underset{\sim}{r}_p: \mathcal{B}_p \rightarrow \mathcal{B}_q \tag{II-4}$$

is a material isomorphism, the elastic response functions of p,q relative to $\underset{\sim}{r}_p$, $\underset{\sim}{r}_q$ satisfy (II-2). If (II-2) is valid for some $\underset{\sim}{r}_p$ and $\underset{\sim}{r}_q$ then (II-4) defines a material isomorphism $\underset{\sim}{r}(p,q)$ of $\mathcal{B}_p$ and $\mathcal{B}_q$.

We call a simple elastic body materially uniform if it consists of pairwise materially isomorphic particles. Thus, in a materially uniform simple elastic body, each particle responds exactly in the same way as every other particle, to a given deformation, provided neighborhoods of these particles are first brought into configurations specified by local reference configurations whose (pairwise) compositions, a la (II-4) comprise material isomorphisms of the respective tangent spaces at the particles (points). In this sense different particles of the body are comprised of the same material, i.e., if its particles are pairwise materially isomorphic, the body is materially uniform.

If $\mathcal{B}$ is materially uniform we define a reference chart for $\mathcal{B}$ to be a pair $(\mathcal{U}_\alpha, \underset{\sim}{r}^\alpha)$ where $\mathcal{U}_\alpha \subset \mathcal{B}$ is an open set and $\underset{\sim}{r}^\alpha$ is a smooth field of local reference configurations on $\mathcal{U}_\alpha$, which has the property that relative to $\underset{\sim}{r}_p$, p$\epsilon\mathcal{U}_\alpha$, the elastic response functions $\underset{\sim}{S}$ are independent of p, i.e., there exists a tensor function $\underset{\sim}{S}_\alpha$ such that

$\underset{\sim}{S}_\alpha(\underset{\sim}{F}) = \underset{\sim}{S}(F,p)$, $\forall p \in \mathcal{U}_\alpha$. We call $\mathcal{U}_\alpha$ a reference neighborhood, $\underset{\sim}{r}^\alpha$ a reference map, and $\underset{\sim}{S}_\alpha$ the response function relative to $(\mathcal{U}_\alpha, \underset{\sim}{r}^\alpha)$. If $\underset{\sim}{r}^\alpha$ is a reference map then $\underset{\sim}{r}^\alpha(p,q) \equiv \underset{\sim}{r}_q^{\alpha-1} \circ \underset{\sim}{r}_p^\alpha$ must be a material isomorphism of p and q for all $p,q \in \mathcal{U}_\alpha$; we also note that there does not necessarily exist a configuration ψ: $\mathcal{U}_\alpha \Rightarrow R^3$ such that $\psi_{*p} = \underset{\sim}{r}_p^\alpha$, $p \in \mathcal{U}_\alpha$.

Now let $(\mathcal{U}_\alpha, \underset{\sim}{r}^\alpha)$, $(\mathcal{U}_\alpha, \underset{\sim}{r}^\beta)$ be two reference charts on $\mathcal{B}$; we call these charts compatible if the induced response functions $\underset{\sim}{S}_\alpha$ and $\underset{\sim}{S}_\beta$ are identical on $\mathcal{U}_\alpha \cap \mathcal{U}_\beta$ and term a collection $\underset{\sim}{\Phi} = \{(\mathcal{U}_\alpha, \underset{\sim}{r}^\alpha), \alpha \in I\}$ of mutually compatible reference charts a reference atlas of $\mathcal{B}$ provided this collection is maximal and $\{\mathcal{U}_\alpha, \alpha \in I\}$ is an open covering of $\mathcal{B}$. It is trivial to see then that if $\underset{\sim}{\Phi} = \{(\mathcal{U}_\alpha, \underset{\sim}{r}^\alpha), \alpha \in I\}$ is a reference atlas for $\mathcal{B}$ the induced response functions $\underset{\sim}{S}_\alpha$ are independent of α, $\alpha \in I$, so we will set $\underset{\sim}{S}_\alpha = \underset{\sim}{S}_{\underset{\sim}{\Phi}}$ and will call $\underset{\sim}{S}_{\underset{\sim}{\Phi}}$ the response function relative to $\underset{\sim}{\Phi}$. When $\mathcal{B}$ possesses a reference atlas $\underset{\sim}{\Phi}$ we will call it a smooth materially uniform (simple) elastic body. In the original work of Noll [5], [6] it was required that $\mathcal{B}$ be covered by a single reference neighborhood $\mathcal{U}$, i.e., that $\mathcal{B}$ have a reference atlas of the form $\underset{\sim}{\Phi}^* = \{(\mathcal{U}, \underset{\sim}{r})\}$; the more general definition of reference atlas given above is due to Wang [7] and brings out the important fact that the smoothness of the field of response functions is, in general, a local and not a global property on $\mathcal{B}$.

Remark. Reference atlases on $\mathcal{B}$ are not unique. It can be shown (i.e., Noll [6], Wang [7]) that if $\underset{\sim}{\Phi}$ is a reference atlas on $\mathcal{B}$ so is $\underset{\sim}{K}\underset{\sim}{\Phi} = \{(\mathcal{U}_\alpha, \underset{\sim}{K}\circ\underset{\sim}{r}^\alpha), \alpha\varepsilon I\}$ where $\underset{\sim}{K}: R^3 \to R^3$ is an isomorphism satisfying det $\underset{\sim}{K}>0$; furthermore, all reference atlases of $\mathcal{B}$ are related in this way by some $\underset{\sim}{K}$. There may also exist reference isomorphisms of $\underset{\sim}{\Phi}$, i.e., isomorphisms of R^3 which satisfy $\underset{\sim}{\Phi} = \underset{\sim}{G}\underset{\sim}{\Phi}$ and the collection of all such reference isomorphisms is identical with the isotropy group $G(\underset{\sim}{\Phi})$ relative to $\underset{\sim}{\Phi}$, which is defined below; this follows from the simple relation

$$\underset{\sim}{S}_{\underset{\sim}{\Phi}}(\underset{\sim}{F}) \equiv \underset{\sim}{S}_{\underset{\sim}{\Phi}}(\underset{\sim}{F}\underset{\sim}{G}), \quad \forall \underset{\sim}{F}\varepsilon GL(3)$$

which characterizes such isomorphisms.

5. The Symmetry Groups of a Materially Uniform Simple Elastic Body.

Material bodies in continuum mechanics are classified according to (i) their defining constitutive equations, which must satisfy the principle of material frame-indifference and (ii) the symmetry (or isotropy) groups which these constitutive equations admit. The first truly rigorous formulation of the concept of the isotropy group in continuum mechanics was given by Noll [28], [29], and the definitions we give below are due, essentially, to him.

Definition II-4 Let $p\varepsilon\mathcal{B}$, a simple elastic body. Then an isomorphism $\underset{\sim}{i}: \mathcal{B}_p \to \mathcal{B}_q$ is a member of the isotropy group

$\mathcal{G}(p)$ iff

$$\underset{\sim}{E}(\phi_{*p}, p) \equiv \underset{\sim}{E}(\phi_{*p} \circ \underset{\sim}{i}, p) \qquad \text{(II-5)}$$

for all configurations $\phi: \mathcal{B} \to R^3$, i.e., iff $\underset{\sim}{i}$ is a material isomorphism of p with itself.

We require (Noll [29]) that $\mathcal{G}(p)$ be a subgroup of $SL(\mathcal{B}_p)$, the special linear group on $\mathcal{B}_p$; $SL(\mathcal{B}_p)$ is that subgroup of $GL(\mathcal{B}_p)$ consisting of isomorphisms $\underset{\sim}{i}$ of $\mathcal{B}_p$ with determinant equal to one. We also make

Definition II-5 An isomorphism $\underset{\sim}{G}: R^3 \to R^3$ belongs to $G(p)$, the isotropy group relative to the local reference configuration $\underset{\sim}{r}_p$ of $p \varepsilon \mathcal{B}$ if for all $F \varepsilon GL(3)$

$$\underset{\sim}{S}(\underset{\sim}{F}, p) = \underset{\sim}{S}(\underset{\sim}{F}\underset{\sim}{G}, p) \qquad \text{(II-6)}$$

From these last two definitions, and the relationships which exist between the response functions $\underset{\sim}{E}$ and $\underset{\sim}{S}$, it is a straightforward matter to deduce (Noll, [6]) that the isotropy groups $G(p)$ and $G(q)$, $p,q \varepsilon \mathcal{B}$ are related via

$$G(p) = \underset{\sim}{r}_p \circ \mathcal{G}(p) \circ \underset{\sim}{r}_p^{-1} \qquad \text{(II-7)}$$

and that

$$\mathcal{G}(q) = \underset{\sim}{r}(p,q) \circ \mathcal{G}(p) \circ \underset{\sim}{r}(p,q)^{-1} \qquad \text{(II-8)}$$

if $\underset{\sim}{r}(p,q): \mathcal{B}_p \to \mathcal{B}_q$ is a material isomorphism. If $\underset{\sim}{r}_p^{\alpha}$ and $\underset{\sim}{r}_p^{\beta}$ are both local reference configurations of $p \varepsilon \mathcal{B}$ and $\hat{\underset{\sim}{G}}: R^3 \to R^3$ is defined by $\hat{\underset{\sim}{G}} = r_p^{\beta} \circ r_p^{\alpha-1}$ then $G_{\alpha}(p)$ and $G_{\beta}(p)$, the isotropy groups relative to these local reference configurations satisfy

$$G_\beta(p) = \hat{\underset{\sim}{G}}G_\alpha(p)\hat{\underset{\sim}{G}}^{-1} \tag{II-9}$$

Also, if $\underset{\sim}{r}(p,q) = \underset{\sim}{r}_q \circ \underset{\sim}{r}_p^{-1}$ is a material isomorphism then (II-7) and (II-8) easily yield $G(p) = G(q)$ so that, in particular, the isotropy groups $G_\alpha(p)$, relative to a reference $\underset{\sim}{r}^\alpha$ on $U_\alpha \subset B$, are independent of p. Following Wang [7] we denote this isotropy group by G_α and call it the isotropy group relative to $(U_\alpha, \underset{\sim}{r}^\alpha)$. From (II-6) it follows that $\underset{\sim}{G} \in G_\alpha$ iff $\underset{\sim}{S}_\alpha(\underset{\sim}{F}) = \underset{\sim}{S}_\alpha(\underset{\sim}{F}\underset{\sim}{G})$ for all $\underset{\sim}{F} \in GL(3)$. If $(U_\alpha, \underset{\sim}{r}^\alpha) \in \Phi$, a reference atlas, then the G_α are independent of α and, in this case, we set $G_\alpha \equiv G(\Phi)$, which we call the isotropy group relative to Φ. Clearly, $\underset{\sim}{G} \in G(\Phi)$ iff $\underset{\sim}{S}_\Phi(\underset{\sim}{F}) = \underset{\sim}{S}_\Phi(\underset{\sim}{F}\underset{\sim}{G})$ for all $\underset{\sim}{F} \in GL(3)$; this confirms our remarks regarding reference isomorphisms which appear at the end of §3. It is not too difficult to show now that the isotropy groups $G(\Phi)$, introduced above, must satisfy the transformation law $G(\underset{\sim}{K}\Phi) = \underset{\sim}{K}G(\Phi)\underset{\sim}{K}^{-1}$ where K is an orientation preserving isomorphism of R^3.

Remark If U_α, U_β are two reference neighborhoods with associated reference maps $\underset{\sim}{r}^\alpha$, $\underset{\sim}{r}^\beta$ then on the overlaps the fields

$$\underset{\sim}{G}_{\alpha\beta}(p) = \underset{p}{r}^\alpha \circ \underset{p}{r}^{\beta-1}, \quad p \in U_\alpha \tag{II-10}$$

are smooth and furthermore, $\underset{\sim}{G}_{\alpha\beta} : U_\alpha \to G(\Phi)$. To see this recall that the compatability of the charts $(\mathcal{U}_\alpha, \underset{\sim}{r}^\alpha)$, $(\mathcal{U}_\beta, \underset{\sim}{r}^\beta)$ in $\underset{\sim}{\Phi}$ implies that

$$\underset{\sim}{E}(\underset{\sim}{F}\circ\underset{\sim}{r}_p^\alpha, p) = \underset{\sim}{E}(\underset{\sim}{F}\circ\underset{\sim}{r}_p^\beta, p), \ p\varepsilon\mathcal{U}_\alpha\cap\mathcal{U}_\beta$$

for all $\underset{\sim}{F}\varepsilon GL(3)$; replacing $\underset{\sim}{F} \to \underset{\sim}{F}\underset{\sim}{G}_{\alpha\beta}^{-1}$ we get

$$\begin{aligned}\underset{\sim}{E}(\underset{\sim}{F}\underset{\sim}{G}_{\alpha\beta}^{-1}\underset{\sim}{r}_p^\alpha, p) &= \underset{\sim}{E}(\underset{\sim}{F}\underset{\sim}{r}_p^\beta; p)\\ &= \underset{\sim}{S}_{\underset{\sim}{\Phi}}(\underset{\sim}{F})\\ &= \underset{\sim}{E}(\underset{\sim}{F}\underset{\sim}{G}_{\alpha\beta}^{-1}\underset{\sim}{r}_p^\beta, p)\\ &= \underset{\sim}{S}_{\underset{\sim}{\Phi}}(\underset{\sim}{F}\underset{\sim}{G}_{\alpha\beta}^{-1})\end{aligned}$$

thus $\underset{\sim}{G}_{\alpha\beta}^{-1}(p)\varepsilon\mathcal{G}(\underset{\sim}{\Phi})$ and, hence, so does $\underset{\sim}{G}_{\alpha\beta}(p)$. Obviously $\underset{\sim}{G}_{\alpha\alpha}(p) = \underset{\sim}{I}$, $\forall p\varepsilon\mathcal{U}_\alpha$ and it is also trivial to show that $\underset{\sim}{G}_{\alpha\beta}(p) = \underset{\sim}{G}_{\beta\alpha}(p)^{-1}$, $\forall p\varepsilon\mathcal{U}_\alpha\cap\mathcal{U}_\beta$, and $\underset{\sim}{G}_{\alpha\beta}(p)\underset{\sim}{G}_{\beta\gamma}(p) = \underset{\sim}{G}_{\alpha\gamma}(p)$, $\forall p\varepsilon\mathcal{U}_\alpha\cap\mathcal{U}_\beta\cap\mathcal{U}_\gamma$. The fields $\underset{\sim}{G}_{\alpha\beta}$ will appear again as the coordinate transformations on the <u>material tangent bundle</u> $\mathcal{T}(\mathcal{B},\underset{\sim}{\Phi})$ and the <u>bundle of reference frames</u> $\mathcal{E}(\mathcal{B},\underset{\sim}{\Phi})$ which we will introduce in §4.

<u>Remark</u> Certain particular types of symmetry will be of interest to us in this chapter as well as in the chapters to follow. Once again, the basic definitions here are due to Noll ([28] and [29]).

<u>Definition II-6</u> A simple elastic particle p is called a (i) <u>solid particle</u>, if there exists a local reference configuration $\underset{\sim}{r}_p$ of p such that $\mathcal{G}(p)\subset\mathcal{O}(3)$, the <u>orthogonal</u>

group over R^3, and

(ii) a fluid crystal particle, if it is a non-solid particle

Special subclasses of the above categories, i.e., simple elastic fluid particles and isotropic solid particles, will be considered as they arise. For all these types of simple elastic particles the isotropy groups are closed Lie subgroups of SL(3); that this is so follows from the fact that every closed subgroup of SL(3) is a Lie subgroup and that the isotropy group $G(\underset{\sim}{\Phi})$ is closed (Wang [7]) if the response function $\underset{\sim}{S}_{\underset{\sim}{\Phi}}$ satisfies the continuity condition

$$\lim_{n\to\infty} \underset{\sim}{S}_{\underset{\sim}{\Phi}}(\underset{\sim}{F}\underset{\sim}{G}_n) = \underset{\sim}{S}_{\underset{\sim}{\Phi}}(\underset{\sim}{F}\underset{\sim}{G}) \qquad \text{(II-11)}$$

for all $\underset{\sim}{F}\varepsilon GL(3)$ and every convergent sequence $\{\underset{\sim}{G}_n\}\subset GL(3)$ such that $\lim_{n\to\infty} \underset{\sim}{G}_n = \underset{\sim}{G}$.

6. Material Charts and Material Atlases. The Material Tangent Bundle $T(B,\underset{\sim}{\Phi})$ and the Bundle of Reference Frames $E(B,\underset{\sim}{\Phi})$.

For an arbitrary differentiable manifold M, the way in which the tangent spaces fit together is characterized by the tangent bundle $T(M)$. If we now specify that $M \equiv B$ (a simple elastic body) then we note that, as $T(B)$ is defined independently of the distribution of the elastic response functions on B, it can not possibly characterize the material structure of the body; in order to represent the material structure of B in differential geometric terms we follow

Wang [7] and introduce the concepts of material chart, material atlas, and material tangent bundle. In all that follows B is taken to be a materially uniform simple elastic body.

Definition II-7 (Wang [7]) A bundle chart $(\mathcal{U}_\alpha, \phi_\alpha)$ of $T(B)$ is called a material chart if the transformations

$$\underset{\sim}{r}_\alpha(p,q) \equiv \phi_{\alpha,q} \circ \phi_{\alpha,p}^{-1} : B_p \rightarrow B_q \tag{II-12}$$

are material isomorphisms $\forall p,q \in \mathcal{U}_\alpha$. Two material charts $(\mathcal{U}_\alpha, \phi_\beta)$, $(\mathcal{U}_\beta, \phi_\beta)$ are termed compatible if $r_{\alpha\beta}(p,q) \equiv \phi_{\alpha,p} \circ \phi_{\beta,q}^{-1} : B_q \rightarrow B_p$ is a material isomorphism for all $p \in \mathcal{U}_\alpha$ and $q \in \mathcal{U}_\beta$. A maximal collection of pairwise compatible material charts is a material atlas of B.

From the above definition we see that the bundle charts in a material atlas are tied together via the rule of material isomorphism. Also, if $(\mathcal{U}_\alpha, \phi_\alpha)$ is a material chart in a material atlas of B and we set

$$\phi_{\alpha,p}^{-1} \equiv \underset{\sim}{r}_p^\alpha : B_p \rightarrow R^3 \tag{II-13}$$

then $\underset{\sim}{r}^\alpha$ is a smooth field of local configurations defined on $\mathcal{U}_\alpha$ and, furthermore,

$$\underset{\sim}{r}_q^{\alpha-1} \circ \underset{\sim}{r}_p^\alpha = \phi_{\alpha,q} \circ \phi_{\alpha,p}^{-1} \equiv \underset{\sim}{r}_\alpha(p,q)$$

is a material isomorphism of p and q. Therefore, $(\mathcal{U}_\alpha, \underset{\sim}{r}^\alpha)$

is a reference chart on $\mathcal{B}$ when $\underset{\sim}{r}^{\alpha}$ is defined by (II-13) and the converse is also easily seen to be valid. It is a simple task to demonstrate that the definition given above for the compatibility of two material charts within a material atlas yields, for the induced reference charts on $\mathcal{B}$, the previous definition for compatibility of charts within a reference atlas, i.e., the fields

$$\underset{\sim}{G}_{\alpha\beta}(p) = \phi_{\alpha,p}^{-1} \circ \phi_{\beta,p} \equiv \underset{\sim}{r}_{p}^{\alpha} \circ \underset{\sim}{r}_{p}^{\beta-1}$$

are smooth and take their values in $G(\underset{\sim}{\Phi})$ where $\underset{\sim}{\Phi} = \{(\mathcal{U}_{\alpha}, \underset{\sim}{r}^{\alpha}), \alpha\varepsilon I\}$; note that $G(\underset{\sim}{\Phi})$, which is the isotropy group of $\mathcal{B}$ relative to $\underset{\sim}{\Phi}$, is a closed Lie subgroup of GL(3), the structure group of $T(\mathcal{B})$. If $\underset{\sim}{\Phi}$ is a given reference atlas for $\mathcal{B}$, the material atlas which corresponds to $\underset{\sim}{\Phi}$ (i.e., the one generated via (II-13) is denoted by $\phi(\underset{\sim}{\Phi})$. The coordinate transformations for $\phi(\underset{\sim}{\Phi})$ are given by the $\underset{\sim}{G}_{\alpha\beta}$ above and the sub-bundle of $T(\mathcal{B})$ whose atlas is taken to be $\phi(\underset{\sim}{\Phi})$ and whose structure group is $G(\underset{\sim}{\Phi})$ is now denoted by $T(\mathcal{B},\underset{\sim}{\Phi})$ and is called the <u>material tangent bundle</u> of $\mathcal{B}$ relative to $\underset{\sim}{\Phi}$. We say that we have "reduced" the bundle $T(\mathcal{B})$ to the bundle $T(\mathcal{B},\underset{\sim}{\Phi})$. As $\mathcal{B}$ is materially uniform it admits a reference atlas $\underset{\sim}{\Phi}$ and then (II-13) sets up a 1-1 correspondence between reference and material atlases on $\mathcal{B}$, i.e., every materially uniform simple elastic body admits a well-defined material atlas, say $\phi(\underset{\sim}{\Phi}) \leftrightarrow \underset{\sim}{\Phi}$, so that the material tangent bundle $T(\mathcal{B},\underset{\sim}{\Phi})$ of $\mathcal{B}$ relative to $\underset{\sim}{\Phi}$ is also well-defined.

Exercise 1 Using the relationship which exists among all reference atlases for B derive the relationship that holds among all corresponding material atlases.

Recall now that a principal fibre bundle is a fibre bundle whose fibre space coincides with its structure group, which then acts on itself via the operation of left-multiplication. Also, as indicated in Chapter I, to each fibre bundle there is associated a principal bundle whose base space, structure group, and coordinate transformations coincide withe those of the given fibre bundle. In particular, we indicated in Chapter I how the bundle of linear frames $E(B)$, which is the associated principal bundle of $T(B)$, could be constructed; using precisely the same type of construction as presented there we can form the bundle of reference frames $E(B,\underset{\sim}{\Phi})$ (relative to a reference atlas $\underset{\sim}{\Phi}$ of B) as the associated principal bundle of $T(B,\underset{\sim}{\Phi})$, as follows: first of all, we define a linear frame $\underset{\sim}{e}_p$ to be a reference frame at p relative to $\underset{\sim}{\Phi}$ if there exists a material chart $(U_\alpha,\phi_\alpha)\varepsilon\phi(\underset{\sim}{\Phi})$, such that $e_p = e_p(\alpha) = \phi_{\alpha,p}(\underset{\sim}{i})$ where $\underset{\sim}{i}$ is the frame consisting of the standard basis vectors of R^3. We denote the set of all reference frames at p relative to $\underset{\sim}{\Phi}$ by $E_p(\underset{\sim}{\Phi})$ and take $E(B,\underset{\sim}{\Phi}) = \bigcup_{p\varepsilon B} E_p(\underset{\sim}{\Phi})$. The projection map π is, of course, a map $\pi: E(B,\underset{\sim}{\Phi}) \rightarrow B$ such that $\pi^{-1}(p) = E_p(\underset{\sim}{\Phi})$, i.e., $E_p(\underset{\sim}{\Phi})$ is the fibre of $E(B,\underset{\sim}{\Phi})$ at p. We now define the bundle atlas $\xi(\underset{\sim}{\Phi}) = \{(U_\alpha,\xi_\alpha),\ \alpha\varepsilon I\}$ by $\xi_\alpha(p,G) = R_{\underset{\sim}{G}}(\underset{\sim}{e}_p)$ for all $\underset{\sim}{e}_p \ \varepsilon\ E_p(\underset{\sim}{\Phi})$

and $\underset{\sim}{G} \varepsilon G(\underset{\sim}{\Phi})$. The bundle maps ξ_α have been chosen so that they are isomorphisms of the form

$$\xi_\alpha : U_\alpha \times G(\underset{\sim}{\Phi}) \to \pi^{-1}(U_\alpha) \subset E(B,\underset{\sim}{\Phi})$$

Exercise 2 Show that the frames $e_p(\alpha)$ satisfy the transformation law $\underset{\sim}{e}_p(\beta) = R_{\underset{\sim}{G}_{\alpha\beta}(p)}(\underset{\sim}{e}_p(\alpha))$ for all $p \varepsilon U_\alpha \cap U_\beta$ where[2] $\underset{\sim}{G}_{\alpha\beta}$ denotes the coordinate transformations of the bundle atlas $\phi(\underset{\sim}{\Phi})$. Show that this result and the definition of the bundle maps ξ_α above implies that the coordinate transformations of $\phi(\underset{\sim}{\Phi})$ and $\xi(\underset{\sim}{\Phi})$ are the same.

As with $T(B,\underset{\sim}{\Phi})$ and $T(B)$, it is relatively easy to see that $E(B,\underset{\sim}{\Phi})$ is a subbundle of the bundle of linear frames $E(B)$.

Exercise 3. Let $\underset{\sim}{\Phi}$ be a reference atlas for B and $E(B,\underset{\sim}{\Phi})$ the bundle of reference frames relative to $\underset{\sim}{\Phi}$. If $\underset{\sim}{K} \varepsilon GL(3)$ then show that

$$E(B,\underset{\sim}{K}\underset{\sim}{\Phi}) = R_{\underset{\sim}{K}^{-1}}(E(B,\underset{\sim}{\Phi}))$$

and that $E(B) = \bigcup_{\underset{\sim}{\Phi}} E(B,\underset{\sim}{\Phi})$. From the transformation rule above it follows that the bundle atlases of $E(B,\underset{\sim}{K}\underset{\sim}{\Phi})$ and $E(B,\underset{\sim}{\Phi})$ are related via: $\xi(\underset{\sim}{K}\underset{\sim}{\Phi}) = \{U_\alpha, R_{\underset{\sim}{K}^{-1}}\xi_\alpha C_{\underset{\sim}{K}^{-1}}, \alpha\varepsilon I\}$ where $C_{\underset{\sim}{K}}(\underset{\sim}{G}) = \underset{\sim}{K}\underset{\sim}{G}\underset{\sim}{K}^{-1}$, $\forall \underset{\sim}{G} \varepsilon GL(3)$. Therefore,
$[R_{\underset{\sim}{K}^{-1}} \circ \xi_\alpha \circ C_{\underset{\sim}{K}^{-1}}](p,\underset{\sim}{G}) = R_{\underset{\sim}{K}^{-1}}(\xi_{\alpha,p}(\underset{\sim}{K}^{-1}\underset{\sim}{G}\underset{\sim}{K}))$, $p\varepsilon U_\alpha$, $\underset{\sim}{G}\varepsilon G(\underset{\sim}{\Phi})$.
This, in turn, implies that $\underset{\sim}{K}^{-1}\underset{\sim}{G}\underset{\sim}{K} \varepsilon G(\underset{\sim}{K}\underset{\sim}{\Phi})$, $\underset{\sim}{G}\varepsilon G(\underset{\sim}{\Phi})$.

We will close this section with some observations

(2) $R_{\underset{\sim}{G}}\underset{\sim}{e}_p$ denotes right-multiplication via $\underset{\sim}{G}$.

concerning the Lie algebras of the isotropy groups $G(\Phi)$ and fundamental fields on the bundles $E(B,\Phi)$. First of all, as the isotropy groups $G(\Phi)$ are closed Lie subgroups of SL(3) they are also Lie groups and have, therefore, associated Lie algebras $g(\Phi)$; recalling our definition in Chapter I, the elements of $g(\Phi)$ consist of all left-invariant vector fields on $G(\Phi)$ and the bracket operation on the algebra is defined by $[u,v] = L_u v$, $\forall u,v \in g(\Phi)$, where L denotes the Lie derivative. By the standard representation of $g(\Phi)$, this Lie algebra is isomorphic to the tangent space to $G(\Phi)$ at the identity element of the group. A simple argument (i.e., see Wang [7]) establishes the fact that the Lie algebra $g(\Phi)$ of $G(\Phi)$ is a Lie subalgebra of sl(3), the Lie algebra of SL(3) which, in turn, is a Lie subalgebra of gl(3). As the isotropy groups $G(\Phi)$ satisfy the transformation law: $G(K\Phi) = KG(\Phi)K^{-1}$, $\forall K \varepsilon GL(3)$, the exponential map (refer to Chapter I, i.e., $\exp(GVG^{-1}) = G \exp(V)G^{-1}$ $\forall G \varepsilon G(\Phi)$, $V \varepsilon g(\Phi)$) can be used to show that the Lie algebras $g(\Phi)$ transform as $g(K\Phi) = Kg(\Phi)K^{-1}$ under a change of the reference atlas.

Remark The transformation law for the Lie algebras $g(\Phi)$, which is given above, defines an equivalence relationship on the class of all Lie subalgebras of gl(3) and an equivalence class relative to this relation is then called *a type of Lie subalgebra* (of gl(3)). A classification of the Lie subalgebras of sl(3) and their corresponding (connected) subgroups of SL(3) can be found in Wang [7]. Further discussion of the ways in which Lie algebras may be associated with simple materials may be found in Nono [30] and Belifante [31]; we note, in particular, that Belifante's approach differs from the one taken here in that he uses the torsion tensor of a (curvature-free) material conncection on $\mathcal{B}$ to define a bracket operation (and hence a Lie algebra structure) on the space of smooth tangent vector fields on $\mathcal{B}$.

Finally, as $G(\underset{\sim}{\Phi})$ is the structure group for $E(\mathcal{B},\underset{\sim}{\Phi})$, and $\xi(\underset{\sim}{\Phi}) = \{(\mathcal{U}_\alpha,\xi_\alpha),\ \alpha\varepsilon I\}$ is the bundle atlas, we may follow the lead of Chapter I (where the construction is carried out for $E(\mathcal{B})$) and define the Lie algebra $\bar{\underset{\sim}{g}}(\underset{\sim}{\Phi})$, consisting of the fundamental fields $\bar{\underset{\sim}{v}}$ on $E(\mathcal{B},\underset{\sim}{\Phi})$, via $\bar{\underset{\sim}{v}}_\alpha(p) \equiv \xi_{\alpha,p*}(\underset{\sim}{v})$, for $p \varepsilon \mathcal{U}_\alpha$, where $\underset{\sim}{v}\varepsilon\underset{\sim}{g}(\underset{\sim}{\Phi})$. If $\pi\colon E(\mathcal{B},\underset{\sim}{\Phi})\to\mathcal{B}$, so that $\pi_*\colon E(\mathcal{B},\underset{\sim}{\Phi})_x \to \mathcal{B}_p$, where $\pi(x) = p$, then it follows that $\pi_*(\bar{v}) = 0$, so that $\bar{\underset{\sim}{g}}(\underset{\sim}{\Phi})$ is isomorphic to ker π_*. The collection of fundamental fields $\bar{\underset{\sim}{v}}$ on $E(\mathcal{B},\underset{\sim}{\Phi})$, i.e., $\bar{\underset{\sim}{g}}(\underset{\sim}{\Phi})$ is easily seen to be a Lie subalgebra of $\bar{\underset{\sim}{g}} = gl(3)$, the set of fundamental fields on $E(\mathcal{B})$.

<u>Exercise</u> 4. Show that the fundamental fields $\bar{\underset{\sim}{v}}$ on $E(\mathcal{B},\underset{\sim}{\Phi})$ are independent of the choice of the bundle charts in $\xi(\underset{\sim}{\Phi})$.

<u>Exercise</u> 5. If $\bar{\underset{\sim}{v}}$ is a fundamental field on $E(\mathcal{B},\underset{\sim}{\Phi})$ show that $R_{\underset{\sim}{K}^{-1}*}(\bar{\underset{\sim}{v}})$ is a fundamental field on $E(\mathcal{B},\underset{\sim}{K}\underset{\sim}{\Phi})$, $\forall\underset{\sim}{K}\varepsilon GL(3)$.

7. Material Connections on Simple Elastic Bodies

Let $\underset{\sim}{\Phi}$ be a given reference atlas for $\mathcal{B}$, a materially uniform simple elastic body, and let $E(\mathcal{B},\underset{\sim}{\Phi})$ be the bundle of reference frames relative to $\underset{\sim}{\Phi}$; as before we denote the bundle atlas by $\xi(\underset{\sim}{\Phi}) = \{(\mathcal{U}_\alpha,\xi_\alpha),\ \alpha\varepsilon I\}$ where $\xi_\alpha\colon \mathcal{U}_\alpha \times G(\underset{\sim}{\Phi}) \to \pi^{-1}(\mathcal{U}_\alpha) \subset E(\mathcal{B},\underset{\sim}{\Phi})$.

A material connection on $\mathcal{B}$ is then defined as follows:

Definition II-8 A material connection H on $\mathcal{B}$ is a "G" connection on $E(\mathcal{B},\underset{\sim}{\Phi})$ where $\underset{\sim}{\Phi}$ is any reference atlas for $\mathcal{B}$.

From Chapter I (§5) we see that a "G" connection H on $E(\mathcal{B},\underset{\sim}{\Phi})$ must satisfy the following condition: if $\lambda(t)$ is a smooth curve in $\mathcal{U}_\alpha \subset \mathcal{B}$, $0 \le t \le T$, and $\rho_t\colon E_{\lambda(0)} \to E_{\lambda(t)}$ are the parallel transports along λ relative to H then the maps $\rho_{t,\alpha}\colon G(\underset{\sim}{\Phi}) \to G(\underset{\sim}{\Phi})$ defined by

$$\rho_{t,\alpha} \equiv \xi^{-1}_{\alpha,\lambda(t)} \circ \rho_t \circ \xi_{\alpha,\lambda(0)} \qquad \text{(II-14)}$$

must be elements of the structure group of $E(\mathcal{B},\underset{\sim}{\Phi})$, i.e., of $G(\underset{\sim}{\Phi})$, for each $t\varepsilon[0,T]$. In other words, for a given curve $\lambda\varepsilon\mathcal{U}_\alpha$, $\rho_t(\alpha)$, $0 \le t < T$, must be a smooth curve in $G(\underset{\sim}{\Phi})$ which passes through the identity element of the isotropy group at $t = 0$. This, in turn, implies that $\rho_t(\alpha)$, $0 \le t < T$, is an integral curve of $\underset{\sim}{g}(\underset{\sim}{\Phi})$ the Lie algebra of $G(\underset{\sim}{\Phi})$, i.e., the tangent vector to $\rho_t(\alpha)$ at $t = 0$ must belong to $\underset{\sim}{g}(\underset{\sim}{\Phi})$.

To state this requirement in terms of local coordinates on $\mathcal{B}$, and thus obtain a restriction on the connection symbols of a material connection H on $\mathcal{B}$, let $\phi: \mathcal{B} \to R^3$ be a configuration of such $\mathcal{B}$ that $\phi(p) = (x^1(p), x^2(p), x^3(p))$. If $(\mathcal{U}_\alpha, \xi_\alpha)$ is in $\xi(\underset{\sim}{\Phi})$ we can represent ξ_α by the component form $\underset{\sim}{\xi}_\alpha = \gamma^i_j \, \underset{\sim}{e}_i \otimes \frac{\partial}{\partial x^j}$, where $\{\underset{\sim}{e}_i\} \equiv \underset{\sim}{i}$ is the standard basis for R^3. For the component form of the parallel transports ρ_t along λ from $\lambda(0)$ to $\lambda(t)$ we have

$$\rho_t \equiv \rho^i_j(t) d_{\lambda(0)} x^i \times \left.\frac{\partial}{\partial x^i}\right|_{\lambda(t)}$$

where the functions $\rho^i_j(t)$ must satisfy the equations of parallel transport, i.e.,

$$\dot{\rho}^i_j + \Gamma^{\;i}_{\ell\,k}\, \rho^\ell_j \dot{\lambda}^h = 0, \; i,j = 1,2,3$$

and the initial condition $\rho^i_j(0) = \delta^i_j$; a unique solution $\rho^i_j(t)$ of (II-14) satisfying this initial condition is guranteed to exist by virtue of the standard existence theorem for ordinary differential equations and the assumed smoothness of λ and $\underset{\sim}{\Gamma}$. Finally, we will represent the "induced parallel transports" $\rho_t(\alpha)$ relative to $(\mathcal{U}_\alpha, \xi_\alpha)$ by $\rho_t(\alpha) = \xi^i_j(t)\underset{\sim}{e}_i \otimes \underset{\sim}{e}^i$. Combining the various component representations given above we deduce easily that

$$\xi^i_j(t) = (\gamma^{-1})^i_k(\lambda(t))\rho^k_\ell(t)\gamma^\ell_j(\lambda(0)) \qquad \text{(II-15)}$$

The tangent vector to $\rho_t(\alpha)$, at $t = 0$, thus has the component form $\dot{\xi}^i_j(0)\underset{\sim}{e}_i \otimes \underset{\sim}{e}^j$ where

$$\dot{\xi}^i_j(0) = \left[\frac{\partial(\gamma^{-1})^i_\ell}{\partial x^m}\bigg|_{\lambda(0)} \dot{\lambda}^m(0) + (\gamma^{-1})^i_k(\lambda(0))\dot{\rho}^k_\ell(0) \right] (\gamma^\ell_j)(\lambda(0)) \qquad \text{(II-16)}$$

But, as $\rho^i_j(t)$ is a solution of the equations of parallel transport, i.e., $\dot{\rho}^k_\ell(0) = -\Gamma^k_{\ell m}(\lambda(0))\dot{\lambda}^m(0)$, substitution in (II-16) then yields

$$\dot{\xi}^i_j(0) = (\gamma^\ell_j)(\lambda(0)) \left[\frac{\partial(\gamma^{-1})^i_\ell}{\partial x^m}\bigg|_{\lambda(0)} - \Gamma^k_{\ell m}(\lambda(0))(\gamma^{-1})^i_k(\lambda(0)) \right] \dot{\lambda}^m(0) \qquad \text{(II-17)}$$

Since $\lambda(0)$ and $\dot{\lambda}(0)$ are both arbitrary we have the following major result (Wang [7], [32]):

<u>Theorem</u> II-3. Let $\underset{\sim}{\Phi}$ be a reference basis for B, $G(\underset{\sim}{\Phi})$ the isotropy group relative to $\underset{\sim}{\Phi}$, $E(B,\underset{\sim}{\Phi})$ be the bundle of reference frames relative to $\underset{\sim}{\Phi}$ and $\xi(\underset{\sim}{\Phi}) = \{(U_\alpha, \xi_\alpha),\ \alpha\varepsilon I\}$ the bundle atlas. If the maps $\xi_{\alpha,p} : G(\Phi) \to E_p$ have the representation $\xi_{\alpha,p} = \gamma^i_j \underset{\sim}{e}_i \otimes \frac{\partial}{\partial x^i}\big|_p$ relative to a local coordinate system (x^i) on $U_\alpha \ni p$ then the functions $\Gamma^i_{jk}(p)$ are the connection symbols of a "G" connection H on $E(B,\underset{\sim}{\Phi})$ iff the matrices

$$\left\{ \gamma^\ell_j \left[\frac{\partial(\gamma^{-1})^i_\ell}{\partial x^m} - \Gamma^k_{\ell m}(\gamma^{-1})^i_k \right],\quad m = 1,2,3 \right\}$$

belong to $\underset{\sim}{g}(\underset{\sim}{\Phi})$ at each point $p\varepsilon U_\alpha$.

<u>Remarks</u>. The above proof follows the argument of Wang [32] except that our maps $\xi_{\alpha,p}$ are inverse to the corresponding

ones which he employs. For a different derivation of this same result one could also look up the proof in [7] (which uses the concept of a connection form which has been avoided here). It is not too difficult to show that "G" connections on $E(B,\underset{\sim}{\Phi})$ correspond to those affine connections on $T(B,\underset{\sim}{\Phi})$ which have the property that the parallel transports of the tangent spaces (which are induced by the maps ρ_t) must be material isomorphisms.

Exercise 6. Verify the above statement concerning affine connections on $T(B,\underset{\sim}{\Phi})$.

Remark For an important class of simple elastic materials known as solid crystal bodies the Lie algebra associated with the isotropy group relative to any reference atlas $\underset{\sim}{\Phi}$ is always trivial, i.e., $\underset{\sim}{g}(\underset{\sim}{\Phi}) = \{\underset{\sim}{0}\}$; from Theorem II-3 we see that material connections on such bodies are characterized by the conditions

$$\Gamma^j_{ik} = - \mu^{-1}{}^m_i \frac{\partial \mu^j_m}{\partial x^k} \tag{II-18}$$

where $\underset{\sim}{\mu} = \underset{\sim}{\gamma}^{-1}$. It is a simple matter to compute that $\underset{\sim}{R} = \underset{\sim}{0}$ (the curvature tensor based on the Γ symbols in (II-18)) so that material connections on solid crystal bodies are both unique and completely integrable. In the earlier work of Noll [5], [6], all material connections were defined so as to be completely integrable; a simple elastic body, however, need not admit any completely integrable material connections and some non-trivial examples of Wang's, to this effect, will be presented at the end of this chapter.

A very interesting class of simple elastic bodies which can be equipped with a torsion-free material connection is the class of solid bodies. A simple elastic material $\mathcal{B}$ is said to be a solid body if, for some reference atlas $\underset{\sim}{\Phi}$, $G(\underset{\sim}{\Phi}) \subset O(3)$, the orthogonal group on R^3. It is trivial to see that this later definition is consistent with definition II-6 if we define a solid body to be one which consists of solid particles. A reference atlas $\underset{\sim}{\Phi}$ with the property that $G(\underset{\sim}{\Phi}) \subset O(3)$ is called <u>undistorted</u>. To proceed we need the following

<u>Definition</u> II-9 (Wang [7]). Let $p \in \mathcal{B}$, a simple elastic body. A tensor $\underset{\sim}{t}$ belonging to $T^{r.s}(p)$ is called <u>intrinsic</u> if it is invariant under all transformations induced by elements of $\mathcal{g}(p)$, i.e., by material isomorphisms of p.

If we choose a chart $(\mathcal{U}_\alpha, \underset{\sim}{r}^\alpha)$ in the reference atlas $\underset{\sim}{\Phi}$, such that $p \in \mathcal{U}_\alpha$, and then define $[g_\Phi(p)](\underset{\sim}{u},\underset{\sim}{v}) = \underset{\sim}{r}^\alpha_p(\underset{\sim}{u}) \cdot \underset{\sim}{r}^\alpha_p(\underset{\sim}{v})$ then it is easy to show that $\underset{\sim}{g}_\Phi$ is an intrinsic tensor in $T^{0,2}(p)$ which can be extended so as to yield an intrinsic Riemannian metric on $\mathcal{B}$; we call this metric the <u>induced metric</u> of the undistorted reference atlas $\underset{\sim}{\Phi}$. That $\underset{\sim}{g}_\Phi$ is intrinsic is a direct consequence of its definition and the fact that the values of Euclidean dot products are preserved under transformations induced by elements of $\mathcal{O}(3)$. Conversely, Wang [7] has shown that if a simple elastic body $\mathcal{B}$ possesses an intrinsic Riemannian metric g then $\mathcal{B}$ must actually be a solid body and there must exist an undistorted reference atlas $\underset{\sim}{\Phi}$ such that $g = g_{\underset{\sim}{\Phi}}$.

Remark If $\underset{\sim}{\Phi}$ is an undistorted reference atlas for $\mathcal{B}$ then, obviously, so is $\underset{\sim}{O}\underset{\sim}{\Phi}$ where $\underset{\sim}{O} \in \mathcal{O}(3)$. It is trivial to show then that $g_{\underset{\sim}{\Phi}} = g_{\underset{\sim}{O}\underset{\sim}{\Phi}}$.

Remark It can be shown that all undistorted reference atlases for $\mathcal{B}$ are of the form $\underset{\sim}{K}\underset{\sim}{\Phi}$ where the symmetric positive-definite part of $\underset{\sim}{K}$, in its polar decomposition, commutes with all elements in $\mathcal{G}(\underset{\sim}{\Phi})$. This fact, in combination with the previous remark, leads to the conclusion that all intrinsic Riemannian metrics on $\mathcal{B}$ have the form $g_{\underset{\sim}{U}\underset{\sim}{\Phi}}$ where $\underset{\sim}{U}$ is any symmetric positive definite tensor in GL(3) which commutes with all elements of the isotropy group $\mathcal{G}(\underset{\sim}{\Phi})$.

Now let H be any material connection on $\mathcal{B}$, a solid elastic body, and let g be any intrinsic Riemannian metric on $\mathcal{B}$. As g is intrinsic, and as the parallel transports relative to H are material isomorphisms of the tangent spaces on $\mathcal{B}$, it follows that the covariant derivative of g relative to H must vanish, i.e.,

$$g_{ij|k} = \frac{\partial g_{ij}}{\partial x^k} - \Gamma^l_{ik} g_{lj} - \Gamma^l_{jk} g_{il} = 0 \qquad \text{(II-19)}$$

where the Γ^i_{jk} are the connection symbols of H. If we now restrict H to be torsion free and solve (II-19) (and its variants which are obtained by permuting indices and using the fact that $g_{ij} = g_{ji}$, $\Gamma^k_{ij} = -\Gamma^k_{ji}$) for the connection symbols Γ we find that

$$\Gamma^i_{jk} = \{{}^{\,i}_{jk}\} = \tfrac{1}{2} g^{im}\left(\frac{\partial g_{km}}{\partial x^j} + \frac{\partial g_{jm}}{\partial x^k} - \frac{\partial g_{jk}}{\partial x^m}\right) \qquad \text{(II-20)}$$

where the $\{{}^{\,i}_{jk}\}$ are the classical Christoffel symbols. In other words, if H is torsion-free, it must coincide with the Riemannian connection relative to g. We have thus proved the following

<u>Theorem</u> II-4 (Wang [7]). A solid body $\mathcal{B}$ can be equipped with at most one torsion-free material connection which, if it exists, must coincide with the Riemannian connection relative to any intrinsic Riemannian metric on $\mathcal{B}$.

From the theorem above it follows at once that no material connection on a solid elastic body $\mathcal{B}$ can be torsion free if the Riemannian connection relative to any intrinsic Riemannian metric on $\mathcal{B}$ is not a material connection; the problem which then arises, of course, is to determine for which classes of solid bodies the Riemannian connections relative to intrinsic Riemannian metrics are material connections. If $\mathcal{B}$ is an <u>isotropic solid body</u>, i.e., $\mathcal{G}(\underset{\sim}{\Phi}) = \mathcal{O}(3)$, then it can be shown that the most general type of symmetric positive definite tensor $\underset{\sim}{U}$ which commutes with all elements of $\mathcal{G}(\underset{\sim}{\Phi})$ is of the form $\underset{\sim}{U} = c\underset{\sim}{I}$, where $c>0$; but, this implies that the intrinsic Riemannian metrics on $\mathcal{B}$ are unique up to a constant factor. This, in turn, implies that there is a unique Riemannian connection relative to all intrinsic Riemannian metrics on an isotropic solid body. But $g(p) = \mathcal{O}(\mathcal{B}_p)$, relative to any intrinsic metric on $\mathcal{B}$, and we have, therefore, the following

Theorem II-5 (Wang [7]). The (unique) Riemannian connection associated with the intrinsic Riemannian metrics on an isotropic solid body is a material connection. Moreover, isotropic solid bodies are the only type of solid bodies for which this is the case.

Remark For solid bodies, other than the isotropic ones, the intrinsic Riemannian metrics do not usually give rise to a unique Riemannian connection; by Theorem II-4, therefore, none of the induced Riemannian connections can be a material connection and, in fact, a torsion-free material connection does not exist on $\mathcal{B}$. We will have more to say about this situation in the next section. For now we note only that (II-19) may be solved for Γ^i_{jk} so as to yield the following relationship between the connection symbols of an arbitrary material connection H on $\mathcal{B}$ and the Christoffel symbols of any intrinsic metric g:

$$\Gamma^i_{jk} = \{^{\,i}_{jk}\} + \tfrac{1}{2} g^{il}(T_{jkl} - T_{ljk} + T_{klj})$$

where $\underset{\sim}{T}$ is the torsion tensor of H and $T_{jkl} = T^i_{jk}\, g_{il}$.

Finally, let us note that, in general, if $\underset{\sim}{g}_\Phi$ is an intrinsic Riemannian metric on $\mathcal{B}$, a solid elastic body, the curvature tensor of the Riemannian connection based on $\underset{\sim}{g}_\Phi$ need not vanish, i.e., the Riemannian connection relative to $\underset{\sim}{g}_\Phi$ is not necessarily a flat connection. It is a well-known classical result in differential geometry that, if the curvature tensor of the Riemannian connection based on a metric such as $\underset{\sim}{g}_\Phi$ is the zero tensor, then there must exist a local coordinate system in a neighborhood of any $p \varepsilon \mathcal{B}$ relative to which the components of $\underset{\sim}{g}_\Phi$ reduce to δ_{ij}. In other words, the metric g on $\mathcal{B}$ is <u>locally Euclidean</u> if the Riemannian connection relative to g is flat. We now make the following

<u>Definition</u> II-10 Let $\underset{\sim}{\Phi}$ be an undistorted reference atlas for the solid elastic body $\mathcal{B}$ and $\underset{\sim}{g}_\Phi$ the intrinsic Riemannian metric on $\mathcal{B}$ induced by $\underset{\sim}{\Phi}$. Then $\underset{\sim}{\Phi}$ is called a <u>regular</u> atlas if $\underset{\sim}{g}_\Phi$ is locally Euclidean.

The following characterization of regular (undistorted) reference atlases has been given by Noll in [6]:

<u>Theorem</u> II-6 An undistorted reference atlas $\underset{\sim}{\Phi} = \{(\mathcal{U}_\alpha, \underset{\sim}{r}^\alpha), \alpha \varepsilon I\}$ is regular iff every reference map $\underset{\sim}{r}^\alpha$ on a simply connected reference neighborhood has a representation of the form

$$\underset{\sim}{r}^\alpha_p = \underset{\sim}{Q}_\alpha(p) \circ \psi_{\alpha,p}, \quad \forall p \varepsilon \mathcal{U}_\alpha \tag{II-21}$$

where $\psi_\alpha : \mathcal{U}_\alpha \to R^3$ is a configuration of $\mathcal{U}_\alpha$ and $\underset{\sim}{Q}_\alpha$ is a smooth field of orthogonal tensors on $\psi_\alpha(\mathcal{U}_\alpha)$.

8. Homogeneity, Local-Homogeneity and Material Connections

Once again, we take $\mathcal{B}$ to be a simple elastic body. The definition given below corresponds to the intuitive picture that we have of the body's being either homogeneous (all particles of $\mathcal{B}$ may be brought into one configuration ϕ from which each responds exactly as any other to a given deformation from this configuration) or locally homogeneous (if $p \in \mathcal{B}$, then all particles in some neighborhood $\mathcal{N}_p$ of p can be brought into one configuration such that each particle in $\mathcal{N}_p$ responds exactly as p does to deformations from this configuration.) In mathematical terms we have

Definition II-11

(i) A simple elastic body B is called homogeneous if it can be equipped with a global reference chart $(\mathcal{B}, \underset{\sim}{r})$ with the property that $\underset{\sim}{r}_p = \psi_{*p}$ for some configuration $\psi: \mathcal{B} \to \mathcal{R}^3$ and all $p \in \mathcal{B}$

(ii) A simple elastic body $\mathcal{B}$ is called locally homogeneous if it can be equipped with a reference atlas $\underset{\sim}{\Phi} = \{(\mathcal{U}_\alpha, \underset{\sim}{r}^\alpha), \alpha \in I\}$ with the property that to each $\alpha \in I$, there corresponds a configuration $\psi^\alpha: \mathcal{U}_\alpha \to \mathcal{R}^3$ such that $\underset{\sim}{r}_p = \psi^\alpha_{*p}$ for all $p \in \mathcal{U}_\alpha$.

Remark The definitions given above are equivalent, in view of the relationship between reference atlases $\underset{\sim}{\Phi}$ on $\mathcal{B}$ and their induced material atlases $\phi(\underset{\sim}{\Phi})$, to the ones given by

Wang [7]; these later definitions are presented in terms of material charts belonging to a pre-bundle atlas of $T(B)$. It should be obvious that every homogeneous simple elastic body B is locally homogeneous and that its material tangent bundles are all trivial, i.e., the base space B can be covered by just one coordinate neighborhood of the bundle atlas. Clearly, even if a locally homogeneous body B admits a reference atlas of the form $\underset{\sim}{\Phi} = \{(B, \underset{\sim}{r})\}$ it does not follow that B is homogeneous, i.e., that there exists a configuration $\psi: B \to R^3$ such that $\underset{\sim}{r}_p = \psi_{*p}$ for all $p \in B$. In the work of Noll [5], [6], only simple bodies B which admit global reference atlases were considered.

We now want to demonstrate that it is possible to characterize the homogeneity, local homogeneity, etc., of a materially uniform simple elastic body B, in terms of the kind of material connections which the body admits. Our first result is the following

Theorem II-7 (Wang [7], Noll [6]) If B admits a flat material connection then it is locally homogeneous.

Proof Let H be a material connection(3) on B which corresponds to the reference atlas $\underset{\sim}{\Phi}$; then each horizontal curve in $E(B)$ is contained in $E(B, \underset{\sim}{\Phi})$, the bundle of reference frames relative to $\underset{\sim}{\Phi}$. If $\tilde{\lambda}$ is such a horizontal curve then in terms of local coordinates relative to (U_α, ξ_α) it is determined by the frames

(3) i.e., a "G" connection on $E(B, \underset{\sim}{\Phi})$.

$$\{e^i_{\cdot j}(t)\frac{\partial}{\partial x^i}\bigg|_{\lambda(t)}, \ j = 1,2,3\}$$

where $\pi(\tilde{\lambda}(t)) = \lambda(t) \in \mathcal{U}_\alpha$. The $e^i_{\cdot j}$ must satisfy the equations of parallel transport

$$\dot{e}^i_{\cdot j}(t) + \Gamma^i_{kl}(\lambda(t))e^k_{\cdot j}(t)\dot{\lambda}^l(t) = 0 \qquad \text{(II-22)}$$

and an initial condition, which we may take, without any loss of generality, to be $e^i_{\cdot j}(0) = \delta^i_{\cdot j}$; the Γ^l_{jk} are, of course, the connection symbols of H. Now, if H is a flat connection then we may choose a local coordinate system $(\bar{x}^i)$ in which $\bar{\Gamma}^i_{jk} \equiv 0$; it then follows that $\{\frac{\partial}{\partial \bar{x}^i}\bigg|_{\lambda(t)}, \ i = 1,2,3\}$, the natural frames of the local coordinate system $(\bar{x}^i)$, are reference frames. But, by definition, a frame $\underset{\sim}{e}_p(\alpha)$, at $p \in \mathcal{U}_\alpha$, is a reference frame iff there exists a material chart $(\mathcal{U}_\alpha, \phi_\alpha) \in \phi(\underset{\sim}{\Phi})$ such that $\underset{\sim}{e}_p(\alpha) = \phi_{\alpha,p}(\underset{\sim}{i})$ or, equivalently, a reference chart $(\mathcal{U}_\alpha, \underset{\sim}{r}^\alpha) \in \underset{\sim}{\Phi}$ such that $e_p(\alpha) = \underset{\sim}{r}_p^{\alpha-1}(\underset{\sim}{i})$. It follows immediately that $\underset{\sim}{r}^\alpha_p = \psi^\alpha_{*p}$ where $\psi^\alpha\colon \mathcal{U}_\alpha \to R^3$ induces the local coordinates $\bar{x}^i$ on $\mathcal{U}_\alpha$. Q.ξ.D.

Remark While the converse of the above theorem is, in general, false a partial converse does exist, i.e., it can be shown that (Wang [7]) if a simple elastic body $\mathcal{B}$ is localy homogeneous then for every $p \in \mathcal{B}$ there exists a material connection H(p) such that the connection symbols of H(p) vanish in some local coordinate system near p.

Recall now that a "G" connection on $E(B)$ is flat iff both the curvature and torsion tensors associated with the connection vanish. Also, as is evident from Theorem II-7 and the above remark, the existence of a flat "G" connection on $E(B,\underset{\sim}{\Phi})$ is related to the local homogeneity of B. Thus, we may use the torsion and curvature tensors associated with material connections on B to characterize the local inhomogeneity of B. Actually, for solid crystal bodies, we may use the torsion tensor of the unique completely integrable material connection, whose connection symbols are given via II-18, as a "measure" of the <u>local dislocation density</u>. In fact, in most of those theories of continuous distributions of dislocations in crystal lattices, where the geometry was prescribed a priori, with no reference to constitutive equations, the prescribed connection was always completely integrable and the associated torsion tensor was always used to characterize the density of the dislocation distribution.

On the other hand, Wang [7] has proven that a simple elastic body B can be equipped with a torsion-free material connection iff there exists a torsion-free material connection $H(U)$ on some neighborhood U of $p \in B$, for each $p \in B$. It then follows from the remark above that every locally-homogeneous simple elastic body can be equipped with a torsion-free material connection. We can summarize our results in the scheme shown below

$\exists$ flat material connection => $\mathcal{B}$ is locally homogeneous

$\Downarrow$

$\exists$ torison-free material connection

Now, if $\mathcal{B}$ is a solid body, then a genuine converse to Theorem II-7 exists and, in fact, we may state

Theorem II-8 A solid elastic body $\mathcal{B}$ is locally homogeneous iff it can be equipped with a flat material connection H.

Proof: the first·part of the theorem is just a restatement of Theorem II-7 so we may go directly to the proof of necessity. As $\mathcal{B}$ is locally homogeneous we know that it can be equipped with a torsion-free material connection; however, as $\mathcal{B}$ is a solid body, at most one such torsion-free material connection can be defined on $\mathcal{B}$, namely, the Riemannian connection relative to any intrinsic Riemannian metric g on $\mathcal{B}$. So, let $\underset{\sim}{\Phi} = \{(\mathcal{U}_\alpha, \underset{\sim}{r}^\alpha), \alpha\varepsilon I\}$ be an undistorted reference atlas for $\mathcal{B}$. As $\mathcal{B}$ is locally homogeneous, we can choose $\underset{\sim}{\Phi}$ so that for each $p\varepsilon\mathcal{B}$, $\exists\alpha\varepsilon I$ and $\psi_\alpha: \mathcal{U}_\alpha \to R^3$ such that $p\varepsilon\mathcal{U}_\alpha$ and $\underset{\sim}{r}^\alpha_p = \psi_{\alpha*p}$, $\forall p\varepsilon\mathcal{U}_\alpha$. If we define

$[g_{\underset{\sim}{\Phi}}(p)](\underset{\sim}{u},\underset{\sim}{v}) = \psi_{\alpha*p}(\underset{\sim}{u}) \cdot \psi_{\alpha*p}(\underset{\sim}{v})$, $\forall \underset{\sim}{u},\underset{\sim}{v}\varepsilon\mathcal{B}_p$, then relative to local coordinates (x^i) induced on $\mathcal{U}_\alpha$ by ψ_α, the components of $g_{\underset{\sim}{\Phi}}$ are just δ_{ij}. Thus $\underset{\sim}{\Phi}$ is regular and the induced metric $g_{\underset{\sim}{\Phi}}$ is locally Euclidean, i.e., the Riemannian connection derived from $g_{\underset{\sim}{\Phi}}$ is flat. Q.E.D

Remarks Following Theorem II-5 we indicated that, in general, the intrinsic Riemannian metrics on solid bodies do not give rise to a unique Riemannian connection and thus a torsion-free material connection can not exist. As Wang [7] has indicated, even if all the intrinsic Riemannian metrics on a solid crystalline body do give rise to a unique Riemannian connection, that connection still need not be a material connection. For instance, if $\mathcal{B}$ is a cubic crystal body then, as is the case with isotropic solid bodies, it can be shown that $\underset{\sim}{U} = c\underset{\sim}{I}$, $c>0$, is the most general symmetric positive-definite tensor which commutes with all elements of $G(\underset{\sim}{\Phi})$, where $\underset{\sim}{\Phi}$ is a reference atlas for $\mathcal{B}$; thus, the intrinsic Riemannian metrics on cubic crystal bodies are unique up to a constant scalar factor and this implies that they give rise to a unique Riemannian connection. However, all material connections on solid crystal bodies must be curvature-free (i.e., of the form (II-18)). If our Riemannian connection were a material connection then it would have to be a flat material connection and this, in turn, implies that $\mathcal{B}$ is locally homogeneous. Therefore, even though all infrinsic Riemannian metrics on a cubic crystal body $\mathcal{B}$ give rise to a unique Riemannian connection, that connection can not be a material connection if $\mathcal{B}$ is not locally homogeneous. Some examples illustrating the results of this section may be found in Wang [7], §10-11.

9. Field Equations of Motion

We wish to indicate in this section how exact field equations of motion may be derived for an elastic body which possesses a continuous distribution of dislocations (inhomogeneities). Our derivation, follows essentially, the approach of Wang [7] and the resulting field equations are more general than those first found by Noll [6] (see also the description of Noll's early work in this area which is contained in Truesdell & Noll [5], §34). As has been pointed out in [7], the derivation to be given here employs no approximation or linearization of any kind and is based soley on the response function of the elastic points and the material geometric structure of the body.

Let $\mathcal{B}$ be a smooth materially uniform elastic body and let H be a fixed material connection on $\mathcal{B}$ i.e., a "G" connection on $E(\mathcal{B},\underset{\sim}{\Phi})$ where $\underset{\sim}{\Phi}$ is some reference atlas for $\mathcal{B}$. We introduce a configuration $\phi\colon \mathcal{B} \to R^3$ and assume that it induces a global coordinate system (x^i) on $\mathcal{B}$. Relative to the system (x^i) we take $\Gamma^i_{jk}(\underset{\sim}{x})$ as the connection symbols of H. Recall that the constitutive equation of an elastic point x is of the form

$$\underset{\sim}{T}_S(x) = \underset{\sim}{E}(\phi_{*x}, x)$$

where $\underset{\sim}{T}_S(x)$ is the stress at the particle x in the configuration ϕ. We may also write this in the form

$$\underset{\sim}{T}_S(x) = \underset{\sim}{S}(\underset{\sim}{F}(x))$$

where $\underset{\sim}{S}$ is the elastic response function relative to $\underset{\sim}{\Phi}$. Of course, if $x\varepsilon\mathcal{U}_\alpha$, where $\underset{\sim}{\Phi} = \{(\mathcal{U}_\alpha, \underset{\sim}{r}^\alpha), \alpha\varepsilon I\}.$, then $\underset{\sim}{F}(x) = \phi_{*x} \circ \underset{\sim}{r}_x^{\alpha-1}$. In component form, relative to the coordinate system (x^i), we may drop the S subscript on $\underset{\sim}{T}$ and write

$$T^i_j(x) = S^i_j([F^p_q(x)]) \tag{II-23}$$

<u>Note</u> The mapping $\underset{\sim}{\mu}: \mathcal{U}_\alpha \to \mathcal{E}(\mathcal{B}, \mathcal{U}_\alpha)$ given via

$$\mu(x) = \{F^i_j(x)\frac{\partial}{\partial x^i}\bigg|_x, \ j = 1,2,3\}, \ x\varepsilon\mathcal{U}_\alpha$$

is easily seen to define a cross-section in $\mathcal{E}(\mathcal{B}, \underset{\sim}{\Phi})$ above the coordinate neighborhood $\mathcal{U}_\alpha$.

Now, in component form the Cauchy equations of motion (refer to §2)

$$\text{div } \underset{\sim}{T}_S + \rho \underset{\sim}{b} = \rho \ddot{\underset{\sim}{x}}$$

assume the form

$$\frac{\partial T^i_j}{\partial x^j} + \rho b^i = \rho \ddot{x}^i \tag{II-24}$$

where the b^i are the components of the external body force in the configuration $\phi(\mathcal{B})$ and the $\ddot{x}^i$ are the components of the acceleration field in some motion of $\mathcal{B}$, say, $\phi(t)$. From (II-23), (II-24), we easily obtain

$$H^{il}_{jk} \frac{\partial F^k_l}{\partial x^j} + \rho b^i = \rho \ddot{x}^i \tag{II-25}$$

where $H^{il}_{jk}([F^p_q]) \equiv \partial S^i_j / \partial F^k_l$ is the gradient of $\underset{\sim}{S}$. Now the response function $\underset{\sim}{S}$ satisfies $\underset{\sim}{S}(\underset{\sim}{F}) = \underset{\sim}{S}(\underset{\sim}{F}\underset{\sim}{G})$, $\forall \underset{\sim}{G} \varepsilon \underset{\sim}{G}(\Phi)$, so

$$S^{i}_{\cdot j}([F^{p}_{q}]) = S^{i}_{\cdot j}([F^{p}_{r}G^{r}_{q}]) \qquad \text{(II-26)}$$

If in (II-26) we let $\underset{\sim}{G}(t) \subset \underset{\sim}{G}(\Phi)$ be a curve in the isotropy group which passes through the identity element of the group at t = 0, then $\dot{G}^{i}_{j}(0) \equiv H^{i}_{\cdot j} \varepsilon \underset{\sim}{g}(\underset{\sim}{\Phi})$ and straightforward differentiation of (II-26) yields

$$H^{il}_{\cdot jk}([F^{p}_{q}])F^{k}_{r}H^{r}_{l} = 0, \ \forall \underset{\sim}{F} \varepsilon GL(3) \qquad \text{(II-27)}$$

after setting t = 0. As $[F^{i}_{\cdot j}]$ defines a cross-section in $\underset{\sim}{E}(B,\Phi)$ over $U_{\alpha} \ni x$ we may apply the condition of Theorem II-3 with $\underset{\sim}{\gamma}^{-1}$ replaced by $\underset{\sim}{F}$, i.e., the $\Gamma^{i}_{jk}(x)$ are the connection symbols of a material connection H on B iff the matrices

$$\hat{H}^{r}_{l} \equiv \{F^{r}_{s}{}^{-1}(x)[\frac{\partial F^{s}_{l}}{\partial x^{m}}\Big|_{x} + \Gamma^{s}_{nm}(x)F^{n}_{l}], \ m = 1,2,3\}$$

belong to $\underset{\sim}{g}(\underset{\sim}{\Phi})$ at each point $x \varepsilon U_{\alpha}$. Replacing H by $\hat{H}$ in (II-27), after setting m = j, yields

$$H^{il}_{\cdot jk}([F^{p}_{q}])\frac{\partial F^{k}_{l}}{\partial x^{j}} = - H^{il}_{\cdot jk}([F^{p}_{q}])\Gamma^{k}_{nj}F^{n}_{l} \qquad \text{(II-27)}$$

which, in turn, may be used to reduce (II-25) to

$$- H^{il}_{\cdot jk}([F^{p}_{q}])\Gamma^{k}_{nj}F^{n}_{l} + \rho b^{i} = \rho \ddot{x}^{i} \qquad \text{(II-28)}$$

which are the equations of motion on the coordinate neighborhood U_{α}. It can be shown (i.e., Wang [7'], §9) that the above equations are independent of the coordinate neighborhood $\mathcal{U}_\alpha \ni x$. Of course, as $\phi = \phi(t)$ is a motion, the connection symbols Γ^i_{jk} as well the local deformation gradients F^p_q are time-dependent. To render this time-dependence explicit we may introduce a fixed reference configuration $\kappa: \mathcal{B} \to \mathcal{R}^3$ which induces a global coordinate system (X^A) on $\mathcal{B}$. The motion $\phi(t)$ is then representable in terms of deformation functions of the form $x^i = x^i(\underset{\sim}{X},t)$. The transformation laws for the Γ^i_{jk} and the F^p_q (from $(X^\alpha) \to (x^i)$) are given by

$$\Gamma^i_{jk}(\underset{\sim}{x}) = \overset{\kappa}{\Gamma}{}^A_{BC}(\underset{\sim}{X}) \frac{\partial x^i}{\partial X^A} \frac{\partial X^B}{\partial x^j} \frac{\partial X^C}{\partial x^k} - \frac{\partial^2 x^i}{\partial X^A X^B} \frac{\partial X^A}{\partial x^j} \frac{\partial X^B}{\partial x^k}$$

$$F^p_q(x) = \overset{\kappa}{F}{}^D_q \frac{\partial x^p}{\partial X^D}$$

where $\overset{\kappa}{\underset{\sim}{\Gamma}}$ and $\overset{\kappa}{\underset{\sim}{F}}$ are the representation of $\underset{\sim}{\Gamma}$ and $\underset{\sim}{F}$ in the (X^A) coordinate system and $\underset{\sim}{x} = \underset{\sim}{x}(\underset{\sim}{X},t)$. If we substitute the relations above into (II-28) we get

$$H^{il}_{jk}\left(\frac{\partial^2 x^k}{\partial X^D \partial X^B} \frac{\partial X^B}{\partial x^j} - \overset{\kappa}{\Gamma}{}^A_{Dc} \frac{\partial x^k}{\partial X^A} \frac{\partial X^C}{\partial x^j}\right) \overset{\kappa}{F}{}^D_l + \rho b^i = \rho \ddot{x}^i \qquad \text{(II-29)}$$

where $H^{il}_{jk} = H^{il}_{jk}([\overset{\kappa}{F}{}^D_q \, x^p_{;D}])$. In the special case where $\mathcal{B}$ is homogeneous, the Euclidean connection in $\kappa(\mathcal{B})$ will be a material connection for an appropriate choice of κ. In the (X^A) coordinate system induced by this particular κ we would then have $\overset{\kappa}{\Gamma}{}^A_{BC}(X^D) \equiv 0$ and $\overset{\kappa}{\underset{\sim}{F}} = \underset{\sim}{I}$ (identity map). It then follows that (II-29) reduces to the usual equations of motion for homogeneous elastic bodies, i.e., to

$$H^{iD}_{jk} \frac{\partial^2 x^k}{\partial X^D \partial X^A} \frac{\partial X^A}{\partial x^j} + \rho b^i = \rho \ddot{x}^i \qquad \text{(II-30)}$$

where $H^{iD}_{jk} = H^{iD}_{jk}([x^i_{,A}])$.

<u>Remarks</u> While the equations (II-29) appear to be of a local nature only, i.e., valid on the coordinate neighborhood $\mathcal{U}_\alpha \subset \mathcal{B}$, Wang ([7] §12) has shown that it is possible to re-write them in the global form

$$\tilde{H}^{iD}_{jk}\left(\frac{\partial^2 x^k}{\partial X^D \partial X^B} - \overset{\kappa}{\Gamma}{}^A_{DB} \frac{\partial x^k}{\partial X^A}\right) \frac{\partial X^B}{\partial x^j} + \rho b^i = \rho \ddot{x}^i \qquad \text{(II-31)}$$

where the $\tilde{H}^{iD}_{jk}([x^n_{,A}]) = H^{il}_{jk}([x^n_{,k} \overset{\kappa}{F}{}^K_q])\overset{\kappa}{F}{}^D_l$.

In terms of the components of the <u>Piola</u> - <u>Kirchhoff</u> <u>stress</u> <u>tensor</u> $\overset{\kappa}{T}{}^A_k$, which are related to T^i_j via

$$\overset{\kappa}{T}{}^A_k = J T^l_k \frac{\partial X^A}{\partial x^l}, \quad J = \det[x^i_{,A}]$$

the Cauchy equations of motion can be cast in the form

$$\frac{\partial \overset{\kappa}{T}{}^A_k}{\partial X^A} + \rho_\kappa b^k = \rho_\kappa \ddot{x}^k \qquad \text{(II-32)}$$

where ρ_κ denotes the mass density in $\kappa(\mathcal{B})$. If, following

Wang ([7], §12) we introduce the response function $\underset{\sim}{A}$ which is defined for all $\underset{\sim}{F}\varepsilon GL(3)$ via

$$A^{i}_{\cdot j}([F^{p}_{q}]) \equiv \det[F^{p}_{q}]S^{k}_{\cdot j}([F^{r}_{s}])F^{i^{-1}}_{\cdot k}$$

then it is possible to show that (II-32) leads to the system of equations

$$\tilde{B}^{DF}_{ik}\left(\frac{\partial^{2}x^{k}}{\partial X^{D}\partial X^{F}} - \overset{\kappa}{\Gamma}{}^{A}_{FD}\frac{\partial x^{k}}{\partial X^{A}}\right) - \tilde{A}^{D}_{\cdot i}\overset{\kappa}{T}{}^{A}_{DA} + \rho_{\kappa}b^{i} = \rho_{\kappa}\ddot{x}^{i} \qquad \text{(II-33)}$$

where

$$\tilde{B}^{DF}_{ik}([x^{l}_{,A}]) \equiv \left(\frac{1}{\det \overset{\kappa}{\underset{\sim}{F}}}\right) B^{sl}_{ik}\overset{\kappa}{F}{}^{D}_{s}\overset{\kappa}{F}{}^{F}_{l}$$

$$B^{sl}_{jk} \equiv \partial A^{s}_{\cdot j}/\partial F^{k}_{l}$$

$$\tilde{A}^{D}_{\cdot i}([x^{j}_{,A}]) \equiv \left(\frac{1}{\det \overset{\kappa}{\underset{\sim}{F}}}\right) A^{s}_{\cdot i}\overset{\kappa}{F}{}^{D}_{s}$$

and

$$\overset{\kappa}{T}{}^{A}_{BC} \equiv \overset{\kappa}{\Gamma}{}^{A}_{BC} - \overset{\kappa}{\Gamma}{}^{A}_{CB}$$

are the components of the torsion tensor of H in the co-ordinate system (X^{A}). If $\mathcal{B}$ is a homogeneous elastic body then the system (II-33) may be reduced to

$$B^{DF}_{ik}\frac{\partial^{2}x^{k}}{\partial X^{D}\partial X^{F}} + \rho_{\kappa}b^{i} = \rho_{\kappa}\ddot{x}^{i}$$

by choosing κ to be a homogeneous reference configuration (i.e., if $\underset{\sim}{\Phi} = (\mathcal{B}, \underset{\sim}{r})$ is a global reference atlas for $\mathcal{B}$ then $\underset{\sim}{r}_p = \kappa_{*p}$ for all $p \in \mathcal{B}$). In this case $\overset{\kappa}{\Gamma}{}^A_{BC}(X^D) \equiv 0$ and $\tilde{B}^{DF}_{ik}$ is easily seen to be independent of (X^A).

Chapter III. Generalized Elastic Bodies

1. Introduction

In the previous chapter we presented a formulation of the theory of continuous distributions of dislocations which originated in the works of Noll [6] and Wang [7]; our results, thus far, apply only to static distributions of inhomogeneities in simple elastic bodies. In this chapter we want to demonstrate that our differential-geometric ideas may be extended so as to cover a much wider class of physical materials which we shall call generalized elastic bodies; among the kinds of materials which may be considered as being generalized elastic bodies are the simple elastic materials discussed in Chapter I as well as bodies which are non-simple, i.e., bodies consisting of oriented elastic particles, quasi-elastic particles, and materials with internal state variables. The concepts to be presented here are based on a 1969 paper of C. C. Wang [32] who, in turn, was motivated by recent work on the thermodynamical theory of simple materials to consider an extension of the geometric theory presented in the previous chapter to certain classes of non-simple and non-uniform materials.

2. Index Sets and Generalized Elastic Bodies

Let $\mathcal{B}$ be a material body, i.e., an oriented three-dimensional differentiable manifold, whose points we consider to be particles, and which is endowed with a non-negative scalar measure that we call the mass distribution of the body. Let $\phi(t)$ be a motion of $\mathcal{B}$ and let $p\varepsilon\mathcal{B}$. Then we recall that p is termed a simple elastic particle if there exists a tensor function $\dot{\underset{\sim}{E}}$ such that

$$\underset{\sim}{T}_S(p,t) = \underset{\sim}{E}(\phi_{*p}(t),p).$$

More generally, if $\underset{\sim}{k}_p(t)$ is any local motion of $p\varepsilon\mathcal{B}$ then p is a simple elastic particle if

$$\underset{\sim}{T}_S(p,t) = \underset{\sim}{E}(\underset{\sim}{k}_p(t),p) \equiv \underset{\sim}{E}_p(\underset{\sim}{k}_p(t))$$

Now, in the thermodynamical theory of simple (elastic) bodies the stress at a particle $p\varepsilon\mathcal{B}$ at time $t>0$ is determined not only by local motions $\underset{\sim}{k}_p(t)$ of p but also by various thermodynamical parameters at p such as the temperature and the temperature gradient at p at time t. We then have a constitutive equation of the form

$$\underset{\sim}{T}_S(p,t) = \underset{\sim}{E}_p(\underset{\sim}{k}_p(t),\ \theta(p,t),\ \underset{\sim}{g}(p,t))^{(4)} \qquad \text{(III-1)}$$

to deal with as well as constitutive equations for the internal energy, the heat flux, and the entropy; as in [32] we shall neglect these other quantities and deal solely with the constitutive equation for the stress tensor. In order to abstract from the thermodynamical situated alluded to, and extend our domain of interest from simple to non-simple elastic particles, it is natral to extend our definition of "local configuration of $p\varepsilon\mathcal{B}$" from simply an orientation preserving

(4) θ represents the temperature and g the temperature gradient

isomorphism of $\mathcal{B}_p$ with R^3 to a set $\{\underset{\sim}{k}_p, \theta_1(p),\ldots, \theta_n(p)\}$ where the $\theta_i(p)$ are certain parameters associated with the particle p (examples of what these parameters might be, besides the thermodynamical quantities mentioned above, will be presented at the end of this section). A <u>local motion</u> of $p\varepsilon\mathcal{B}$ is then a one-parameter family of local configurations, i.e., $\{\underset{\sim}{k}_p(t), \theta_1(p,t),\ldots, \theta_n(p,t)\}$. We will call the class of all sets $\{\theta_1(p),\ldots, \theta_n(p)\}$ of physical parameters associated with $p\varepsilon\mathcal{B}$ the <u>index set</u> of p and, following Wang [32], we will denote this set by N_p.(5) If $\mathcal{D}_p$ denotes the class of all orientation-preserving maps of $\mathcal{B}_p$ with R^3, then the constitutive equation of a <u>generalized elastic particle</u> p takes the form

$$\underset{\sim}{T}_S(p,t) = \underset{\sim}{E}_p(\underset{\sim}{k}_p(t), \underset{\sim}{\theta}_p(t)) \tag{III-2}$$

where $\underset{\sim}{k}_p: R^+ \to \mathcal{D}_p$ and $\underset{\sim}{\theta}_p: R^+ \to N_p$. In other words, for each fixed $t\varepsilon R^+$, $\underset{\sim}{E}_p: \mathcal{D}_p \times N_p \to L_s(R^3,R^3)$, where $L_s(R^3,R^3)$ denotes the set set of all symmetric linear maps of R^3 onto R^3.

<u>Remark</u> We will require here that the index set N_p form a finite dimensional differentiable manifold with, say, dim $N_p = n$; for each $t\varepsilon R^+$ the domain of $\underset{\sim}{E}_p$ (the <u>response function</u> of $p\varepsilon\mathcal{B}$) is then the finite dimensional <u>product manifold</u> $\mathcal{D}_p \times N_p$. If we set $N(\mathcal{B}) \equiv \bigcup_{p\varepsilon\mathcal{B}} N_p$ then the map $\underset{\sim}{\theta}_p$ introduced above gives rise to a cross-section $\underset{\sim}{\theta}(t_o)$ of this <u>index bundle</u> $N(\mathcal{B})$ via the map

(5) we have also used N_p to denote a neighborhood of $p\varepsilon\mathcal{B}$ but no confusion should arise here as the meaning will be clear from the context.

$$[\underset{\sim}{\theta}(t_o)](p) \equiv \theta_p(t_o); \quad \forall p \varepsilon \mathcal{B} \qquad \text{(III-3)}$$

we will call $\underset{\sim p}{\theta}(\cdot)$ the <u>index function</u> and $\underset{\sim}{\theta}(t_o)$: $\mathcal{B} \rightarrow \mathcal{N}(\mathcal{B})$ the <u>index section</u>. As already alluded to in the case where the $\underset{\sim i}{\theta}(p)$ are certain thermodynamical parameters associated with the particle $p\varepsilon\mathcal{B}$, the index function $\underset{\sim p}{\theta}(\cdot)$ cannot, in general, be determined by mechanical principles alone and further constitutive equations, beyond that laid down for the stress tensor, must be introduced. We note also that in certain cases, i.e., the oriented elastic particle p, the stress tensor need not be symmetric; however, as Wang [32] indicates, the material geometric structure of the generalized elastic body is insensitive to this fact and thus we will restrict ourselves to response functions $\underset{\sim p}{E}$ which take their values in $\mathcal{L}_s(R^3, R^3)$.

Some examples of specific types of index sets are now in order; the most important ones are

(i) $\mathcal{N}_p$ is either a trivial set consisting of just one point $(\theta_1(p), \ldots, \theta_n(p))$ or the index function θ_p: $R^+ \rightarrow \mathcal{N}_p$ has a constant value. In both cases the constitutive equation for the stress tensor is trivially seen to be that of a <u>simple elastic particle</u> p, i.e., we may define

$$\underset{\sim p}{E}(\underset{\sim p}{k}(t)) \equiv \underset{\sim p}{E}(\underset{\sim p}{k}(t), \theta_1(p), \ldots, \theta_n(p))$$

for any $\underset{\sim p}{k}$: $R^+ \rightarrow \mathcal{D}_p$.

(ii) p is a <u>simple thermoelastic particle</u>, i.e.,

$\underset{\sim}{\theta}_p(t) = (\theta(p,t),\ \underset{\sim}{g}(p,t))$ where $\theta(p,t)$ and $\underset{\sim}{g}(p,t)$ are, respectively, the temperature and temperature gradient, in some local motion $\underset{\sim}{k}_p(t)$ of p, at time t. Thus, $N_p \equiv R^+ \times R^3$ for each $p \varepsilon B$.

(iii) p is a quasi-elastic particle, i.e., the "memory" effect of the particle is represented by an explicit dependence of $\underset{\sim}{E}_p$ on t; the index set here is $N_p \equiv R^+ \times R^3 \times R^+$ and specifically we have

$$\underset{\sim}{\theta}_p(t) = (\theta(p,t),\ \underset{\sim}{g}(p,t),\ t)$$

Quasi-elastic response will be of considerable interest to us in the next chapter where we discuss anelastic behavior and dislocation motion.

(iv) p is an oriented particle, i.e., for $t \varepsilon R^+$, $\underset{\sim}{\theta}_p(t) \ \varepsilon\ R^3$; more general cases can be considered in which N_p is the Cartesian product of R^3 with itself and some tensor spaces over R^3. The value of $\theta_p(t_o)$ at any fixed t_o is called the director of p at t_o.

(v) p is a simple elastic particle with internal variables, i.e., $N_p \equiv R^+ \times R^3 \times R^n$ so that $\underset{\sim}{\theta}_p$ is of the form

$$\underset{\sim}{\theta}_p(t) = (\theta(p,t),\ \underset{\sim}{g}(p,t),\ \theta_1(p,t), \ldots\ \theta_n(p,t))$$

where the $\theta_i(p,t)$ are termed internal variables.

3. Local Material Automorphisms, Phase and Transition Points, the Phase Isotropy Group.

As was the case in the previous chapter, where we dealt with simple elastic particles, it is the concept of material isomorphism which gives rise to the material geometric structures on generalized elastic bodies; for generalized elastic particles however, this concept is slightly more involved than that which was appropriate for the situation encountered in Chapter I. We shall begin with a series of four definitions:

Definition III-1 Let $(\underset{\sim}{\kappa}_o, \theta_o)$ be a fixed point in $\mathcal{D}_p \times \mathcal{N}_p$. A local material automorphism of p at $(\underset{\sim}{\kappa}_o, \theta_o)$ is an automorphism $\underset{\sim}{A}: \mathcal{B}_p \rightarrow \mathcal{B}_p$ such that

$$\underset{\sim}{E}_p(\underset{\sim}{\kappa}, \theta) \equiv \underset{\sim}{E}_p(\underset{\sim}{K} \circ \underset{\sim}{A}, \theta) \tag{III-4}$$

for all $(\underset{\sim}{\kappa}, \theta)$ in some neighborhood $\mathcal{N}(\underset{\sim}{\kappa}_o, \theta_o)$.

Definition III-2 The collection $\mathcal{G}_p(\underset{\sim}{\kappa}_o, \theta_o)$ consisting of all local material automorphisms of p at $(\underset{\sim}{\kappa}_o, \theta)$ is called the local isotropy group of p at $(\underset{\sim}{\kappa}_o, \theta_o)$.

Definition III-3 If the local isotropy group $\mathcal{G}_p(\underset{\sim}{\kappa}_o, \theta_o)$ of p at $(\underset{\sim}{\kappa}_o, \theta_o)$ is invariant on some neighborhood $\mathcal{N}(\underset{\sim}{\kappa}_o, \theta_o)$ then we will call $(\underset{\sim}{\kappa}_o, \theta_o)$ a phase point in $\mathcal{D}_p \times \mathcal{N}_p$; if no such neighborhood exists we say that $(\underset{\sim}{\kappa}_o, \theta_o)$ is a transition point in $\mathcal{D}_p \times \mathcal{N}_p$.

Definition III-4 Two phase points $(\underset{\sim}{\kappa}_o, \theta_o)$ and $(\underset{\sim}{\kappa}_1, \underset{\sim}{\theta}_1)$ in $\mathcal{D}_p \times \mathcal{N}_p$ have the same phase if $G_p(\underset{\sim}{\kappa}_o, \theta_o) = G_p(\underset{\sim}{\kappa}_1, \underset{\sim}{\theta}_1)$. An equivalence class induced by this relation, say $\mathcal{P}_p$, is called a phase of p and $G(\mathcal{P}_p)$, the phase isotropy group of $\mathcal{P}_p$, is then defined to be $G_p(\underset{\sim}{\kappa}_o, \underset{\sim}{\theta}_o)$ for any $(\underset{\sim}{\kappa}_o, \theta_o) \in \mathcal{P}_p$.

Thus, according to the last definition given above, if $p \in \mathcal{B}$ and $(\underset{\sim}{\kappa}_o, \theta_o)$ is any phase point in $\mathcal{D}_p \times \mathcal{N}_p$,

$\mathcal{P}_p \equiv \{(\underset{\sim}{\kappa}, \theta) \in \mathcal{D}_p \times \mathcal{N}_p \mid (\underset{\sim}{\kappa}, \theta)$ is a phase point in $\mathcal{D}_p \times \mathcal{N}_p$ and $G_p(\underset{\sim}{\kappa}, \theta) = G_p(\underset{\sim}{\kappa}_o, \theta_o)\}$

Remarks Using definitions III-1 through III-4 it is a simple exercise to show that any phase $\mathcal{P}_p$ of $p \in \mathcal{B}$ must be an open set in $\mathcal{D}_p \times \mathcal{N}_p$ and that $\mathcal{P}_p$ is closed under the action of the isomorphisms in $G(\mathcal{P}_p)$, i.e., if $(\underset{\sim}{\kappa}_o, \theta_o) \in \mathcal{P}_p$ and $\underset{\sim}{A} \in G(\mathcal{P}_p)$ then $(\underset{\sim}{\kappa}_o \circ \underset{\sim}{A}, \theta) \in \mathcal{P}_p$. As Wang [32] notes it may turn out that for an arbitrary generalized elastic particle $p \in \mathcal{B}$ all points $(\underset{\sim}{\kappa}, \theta) \in \mathcal{D}_p \times \mathcal{N}_p$ are transition points. One type of transition point which can be singled out is called a multiple point; this is a transition point $(\underset{\sim}{\kappa}_o, \theta_o)$ which belongs to $\partial \mathcal{P}_p^1 \cap \ldots \cap \partial \mathcal{P}_p^N$ where $\mathcal{P}_p^1, \ldots, \mathcal{P}_p^N$ are N non-empty phases of p. For a multiple point $(\underset{\sim}{\kappa}_o, \theta_o)$ it can be shown that $G_p(\underset{\sim}{\kappa}_o, \theta_o) \subset G(\mathcal{P}_p^1) \cap \ldots \cap G(\mathcal{P}_p^N)$.

The physical interpretation of (III-4) is clear: it says that an automorphism $\underset{\sim}{A}: \mathcal{B}_p \to \mathcal{B}_p$ is a local material automorphism of p at $(\underset{\sim}{\kappa}_o, \theta_o) \in \mathcal{D}_p \times \mathcal{N}_p$ if the local configurations $\underset{\sim}{\kappa}_o$ and $\underset{\sim}{\kappa}_o \circ \underset{\sim}{A}$ can not be distinguished (via measurement of the stress at p) in any small deformation superimposed on κ_o and any small change of the index θ_o. Note that the condition which would define a <u>material automorphism</u>, say $\underset{\sim}{A}^*$, of p in the present situation is that

$$\underset{\sim}{E}_p(\underset{\sim}{\kappa},\theta) \equiv \underset{\sim}{E}_p(\underset{\sim}{\kappa}\circ\underset{\sim}{A}^*,\theta) \qquad \text{(III-5)}$$

for <u>all</u> $(\underset{\sim}{\kappa},\theta) \in \mathcal{D}_p \times \mathcal{N}_p$ (and not just for all $(\underset{\sim}{\kappa},\theta)$ in some neighborhood of a fixed point $(\underset{\sim}{\kappa}_o,\theta_o) \in \mathcal{D}_p \times \mathcal{N}_p$). In order to have our notation here conform with that introduced in Chapter II, we shall denote the group of all material automorphisms of p by $\tilde{g}(p)$; if the index is fixed, i.e., if (III-5) is required to hold for all $\underset{\sim}{\kappa}\in\mathcal{D}_p$ and a given $\theta\in\mathcal{N}_p$ then $\tilde{g}(p)$ is formally similar to the group $g(p)$ introduced in the previous chapter with the exception that it now varies, in general, with the index θ.

<u>Exercise</u> 7. Show that if $p\in\mathcal{B}$ is a generalized elastic particle then

$$\tilde{g}(p) \subset \bigcup_{(\underset{\sim}{\kappa},\theta)\in\mathcal{D}_p\times\mathcal{N}_p} G_p(\underset{\sim}{\kappa},\theta)$$

<u>Exercise</u> 8 If p has a non-empty phase $\mathcal{P}_p$ prove that the boundary $\partial\mathcal{P}_p$ of $\mathcal{P}_p$ in $\mathcal{D}_p \times \mathcal{N}_p$ must consist soley of transition points

4. Material Isomorphism and Generalized Elasticity

We begin with the following

Definition III-5 Let $p,q \in \mathcal{B}$ be two generalized elastic particles. Then p and q are called materially isomorphic if these exists an orientation preserving isomorphism $\underset{\sim}{r}(p,q)\colon \mathcal{B}_p \to \mathcal{B}_q$ and a diffeomorphism $\underset{\sim}{i}(p,q)\colon \mathcal{N}_p \to \mathcal{N}_q$ such that

$$\underset{\sim}{E}_p(\underset{\sim}{\kappa},\theta) \equiv \underset{\sim}{E}_q(\underset{\sim}{\kappa} \circ \underset{\sim}{r}(p,q)^{-1}, [\underset{\sim}{i}(p,q)](\theta)) \qquad \text{(III-6)}$$

for all $(\underset{\sim}{\kappa},\theta) \in \mathcal{D}_p \times \mathcal{N}_p$.

Using the definition above we can now define a map $\underset{\sim}{I}\colon \mathcal{D}_p \times \mathcal{N}_p \to \mathcal{D}_q \times \mathcal{N}_q$ via $\underset{\sim}{I}(\underset{\sim}{\kappa},\theta) = (\underset{\sim}{\kappa} \circ \underset{\sim}{r}(p,q)^{-1}, [\underset{\sim}{i}(p,q)](\theta))$; we call $\underset{\sim}{I}$ a material isomorphism of p and q if $\underset{\sim}{E}_p = \underset{\sim}{E}_q \circ \underset{\sim}{I}$ on $\mathcal{D}_p \times \mathcal{N}_p$. By using the definition of local material automorphism it is a relatively simple matter to prove the following two results concerning the relationship which exits among the local isotropy groups, on the one hand, and among the phase isotropy groups on the other

Theorem III-1 If $\underset{\sim}{I}\colon \mathcal{D}_p \times \mathcal{N}_p \to \mathcal{D}_q \times \mathcal{N}_q$ is a material isomorphism of the generalized elastic particles $p,q \in \mathcal{B}$ then

$$\mathcal{G}_p(\underset{\sim}{\kappa},\theta) = \underset{\sim}{r}(p,q)^{-1} \circ \mathcal{G}_q(\underset{\sim}{I}(\underset{\sim}{\kappa},\theta)) \circ \underset{\sim}{r}(p,q) \qquad \text{(III-7)}$$

Theorem III-2 A point $(\underset{\sim}{\kappa},\theta) \in \mathcal{D}_p \times \mathcal{N}_p$ is a phase point (transition point) iff $\underset{\sim}{I}(\underset{\sim}{\kappa},\theta) \in \mathcal{D}_q \times \mathcal{N}_q$ is a phase point (transition point) where $I\colon \mathcal{D}_p \times \mathcal{N}_p \to \mathcal{D}_q \times \mathcal{N}_q$ is a material isomorphism of $p,q \in \mathcal{B}$. Also, $\mathcal{P}_p \subset \mathcal{D}_p \times \mathcal{N}_p$ is a phase of p iff $\mathcal{P}_q = I(\mathcal{P}_p) \subset \mathcal{D}_q \times \mathcal{N}_q$ is a phase of q and

$$\mathcal{G}(\mathcal{P}_p) = \underset{\sim}{r}(p,q)^{-1} \circ \mathcal{G}(\mathcal{P}_q) \circ \underset{\sim}{r}(p,q) \qquad \text{(III-8)}$$

If each pair of particles p,q belonging to the generalized elastic body $\mathcal{B}$ are materially isomorphic in the sense of Definition III-5 then we say that $\mathcal{B}$ is a materially uniform generalized elastic body; for such a body we set

$$\mathcal{M}(\mathcal{B}) = \bigcup_{p\in\mathcal{B}} \mathcal{D}_p \times \mathcal{N}_p$$

and define a phase of $\mathcal{B}$ via

$$\mathcal{P}(\mathcal{B}) = \bigcup_{p\in\mathcal{B}} \mathcal{P}_p \subset \mathcal{M}(\mathcal{B})$$

where any two phases $\mathcal{P}_p$ and $\mathcal{P}_q$ in the above union are related via $\mathcal{P}_p = \underset{\sim}{I}(\mathcal{P}_q)$, with $\underset{\sim}{I}$ a material isomorphism of p and q.

Definition III-6 A motion $\phi(t)$: $\mathcal{B} \to R^3$ is said to lie in the phase $\mathcal{P}(\mathcal{B})$ if the induced local motion $(\phi_{*p}(t), \underset{\sim}{\theta}_p(t)) \in \mathcal{P}(\mathcal{B})$, $\forall t \in R^+$ and all $p\in\mathcal{B}$.

As in [32] we shall now restrict our attention to a single phase $\mathcal{P}(\mathcal{B})$ of the materially uniform generalized elastic body $\mathcal{B}$.

Now, let $\mathcal{P}_p$ be the phase of $p\in\mathcal{B}$ which belongs to $\mathcal{P}(\mathcal{B})$. Then, following Wang [32] we can write $\mathcal{P}_p$ as the disjoint union

$$\mathcal{P}_p = \bigcup_{\theta\in\mathcal{V}_p} \mathcal{W}_p(\theta)$$

where $\mathcal{V}_p$ is open in $\mathcal{N}_p$ and where $\mathcal{W}_p(\theta)$ is open in $\mathcal{D}_p$ for each $\theta\in\mathcal{V}_p$. We refer to $\mathcal{V}_p$ as the index range and call $\mathcal{W}_p(\theta)$ the deformation range at θ for the phase $\mathcal{P}_p$.

Definition III-7 A diffeomorphism $\underset{\sim}{\xi}_p: V_p \to V_p$ is called an index automorphism of $p \varepsilon B$ in the phase P_p if for all $(\underset{\sim}{\kappa}, \theta) \varepsilon P_p$.

$$\underset{\sim}{E}_p(\underset{\sim}{\kappa}, \theta) \equiv \underset{\sim}{E}_p(\underset{\sim}{\kappa}, \xi_p(\theta))) \qquad \text{(III-9)}$$

Definition III-8 The collection $I_p(P_p)$ of all index automorphisms $\underset{\sim}{\xi}_p: V_p \to V_p$ is called the index isotropy group of $p \varepsilon B$ in the phase P_p.

Remark In Definition III-5 we presented one concept of material isomorphism which is associated with particles belonging to a generalized elastic body. We may also define a material isomorphism of $p,q \varepsilon B$ with respect to $P(B)$ as an isomorphism $\underset{\sim}{I}: P_p \to P_q$ of the form

$$\underset{\sim}{I}(\underset{\sim}{\kappa}, \theta) = (\underset{\sim}{\kappa} \circ \underset{\sim}{r}(p,q)^{-1}, [\underset{\sim}{i}(p,q)]\theta),$$

where $\underset{\sim}{i}(p,q): V_p \to V_q$ and the condition $\underset{\sim}{E}_p = \underset{\sim}{E}_q \circ \underset{\sim}{I}$ is required to hold on the phase P_p only; such a definition is clearly more restrictive than the one presented in Definition III-5. If $\underset{\sim}{I}$ is a material isomorphism of $p,q \varepsilon B$ with respect to the phase $P(B)$ then it is easy to see that the transformation law (III-8) for the phase istropy group is still valid and, moreover, the index isotropy groups transform according to the rule

$$I(P_p) = \underset{\sim}{i}(p,q)^{-1} \circ I(P_q) \circ \underset{\sim}{i}(p,q) \qquad \text{(III-10)}$$

Exercise 9 Verify the transformation rule for the index isotropy group which is given by III-10.

Remark Recall that in Chapter I we required that $G(\underset{\sim}{\Phi})$, the isotropy group associated with the reference atlas $\underset{\sim}{\Phi}$, be a Lie group. In the present situation we will require that both the phase and index isotropy groups be Lie groups. As the identity map of V_p is the identity element of $I(P_p)$ it is easy to see that the index isotropy group $I(P_p)$ acts as an (effective) Lie transformation group on V_p.

5. The Material - Index Atlas.

We want to begin by introducing the concepts of index fibre space and material fibre space; to this end we make

Definition III-9 Let $p \varepsilon B$ (a generalized elastic body); then, a differentiable manifold V which is diffeomorphic to the index range V_p of any particle $p \varepsilon B$ is called an index fibre space of B with respect to the given phase $P(B)$.

Remarks According to definition III-9, V is an index fibre space of B with respect to $P(B)$ iff there exists, for each $p \varepsilon B$, a diffeomorphism $\underset{\sim}{\eta}_p: V_p \to V$; such a diffeomorphism is called an index reference configuration of p. If B is a materially uniform generalized elastic body, and $\underset{\sim}{I}$ is defined via the remark following Definition III-8, then $\underset{\sim}{\eta}_q \circ \underset{\sim}{i}(p,q): V_p \to V$ is an index reference configuration of p if $\underset{\sim}{\eta}_q$ is an index reference configuration of q. If

$\eta_p\colon V_p \to Y$ is an index reference configuration then

$$I_p \equiv \eta_p \circ I(P_p) \circ \eta_p^{-1}\colon Y \to Y$$

is called the <u>relative index isotropy group</u> of p (with respect to η_p). Clearly, I_p acts as an effective Lie transformation group on the index fibre space.

<u>Definition</u> III-10 Let $p \in B$ (a generalized elastic body). Any diffeomorphism $\zeta_p \in D_p$ is termed a <u>(local) reference configuration</u> of p and R^3, the diffeomorphic copy of B_p for any $p \in B$, is called the <u>material fibre space</u>.

<u>Remark</u> As in Chapter I,

$$G_p \equiv \zeta_p \circ G(P_p) \circ \zeta_p^{-1}\colon R^3 \to R^3$$

is an effective Lie transformation group on R^3; we call G_p the <u>relative phase isotropy group of p</u> with respect to ζ_p

Let η_p, ζ_p be an index reference configuration and a (local) reference configuration of $p \in B$. As in Chapter I we may define the concept of a relative response S_p^* of p which, in this case, is taken with respect to η_p and ζ_p. Our defining equation is

$$S_p^*(F,y) = E_p(F \circ \zeta_p,\ \eta_p^{-1}(y)) \qquad \text{(III-11)}$$

for all $F \in GL(3)$ and all $y \in Y$. We note that while $S_p^*\colon GL(3) \times Y \to L(R^3,R^3)$, we are really interested in the values of S_p^* for $(\underset{\sim}{F},y) \in P_p \equiv \bigcup_{y\in Y} W_p(y)$ where for each $Y = \underset{\sim}{\eta}_p(\theta)$, $W_p(y) = W_p(\theta) \circ \underset{\sim}{\zeta}_p^{-1}$

Exercise 9 Show that

$$S_p^*(\underset{\sim}{F}\underset{\sim}{K},y) \equiv S_p^*(\underset{\sim}{F},y)$$

$$S_p^*(\underset{\sim}{F},\xi(y)) \equiv S_p^*(\underset{\sim}{F},y)$$

for all $\underset{\sim}{F} \in W_p(y)$, $\underset{\sim}{K} \in G_p$, $y\in Y$ and $\xi \in I_p$.

We can now define the concept of a material - index chart in B with respect to the phase $P(B)$; if the phase and index bundles of U_α are denoted respectively by $T(U_\alpha) \equiv {}_{p\in U_\alpha} B_p$ and $V(U_\alpha) \equiv \bigcup_{p\in U_\alpha} V_p$ then we have

Definition III-11 A triple $(U_\alpha, \underset{\sim}{\zeta}_\alpha, \underset{\sim}{\eta}_\alpha)$ is a material-index chart in B - with respect to the phase $P(B)$ - if U_α is an open set in B and

(i) $\underset{\sim}{\zeta}_\alpha : T(U_\alpha) \to U_\alpha \times R^3$ is a diffeomorphism such that $\underset{\sim}{\zeta}_\alpha(B_p) \equiv \{p\} \times R^3$ for each $p\in U_\alpha$

(ii) $\underset{\sim}{\eta}_\alpha : V(U_\alpha) \to U_\alpha \times Y$ is a diffeomorphism such that $\underset{\sim}{\eta}_\alpha(V_p) \equiv \{p\} \times Y$ for each $p\in U_\alpha$

(iii) If $\underset{\sim}{\zeta}_{\alpha,p}$ and $\underset{\sim}{\eta}_{\alpha,p}$ denote the restrictions of $\underset{\sim}{\zeta}_\alpha$ and $\underset{\sim}{\eta}_\alpha$ to B_p and V_p, respectively, and we set

$$\begin{aligned} \hat{\underset{\sim}{r}}(p;q) &\equiv \underset{\sim}{\zeta}_{\alpha,q}^{-1} \circ \underset{\sim}{\zeta}_{\alpha,p} \\ \hat{\underset{\sim}{i}}(p,q) &\equiv \underset{\sim}{\eta}_{\alpha,q}^{-1} \circ \underset{\sim}{\eta}_{\alpha,p} \end{aligned} \qquad \text{(III-13)}$$

then $\underset{\sim}{I}: P_p \to P_q$, as defined in the remark following Definition III-8 with $\underset{\sim}{r} \to \hat{\underset{\sim}{r}}$ and $\underset{\sim}{i} \to \hat{\underset{\sim}{i}}$, must be a material isomorphism of p and q with respect to the phase $P(B)$.

Theorem III-3 If $(U_\alpha, \underset{\sim}{\zeta}_\alpha, \underset{\sim}{\eta}_\alpha)$ is a material-index chart in B with respect to the phase $P(B)$ then S_p^*, taken with respect to the reference configurations $\underset{\sim}{\zeta}_{\alpha,p}: B_p \to R^3$ and $\underset{\sim}{\eta}_{\alpha,p}: V_p \to V$, is independent of p.

Proof The proof follows directly from (III-11) and the fact that $\hat{\underset{\sim}{r}}(p,q)$, $\hat{\underset{\sim}{i}}(p,q)$, as defined in (III-13) give rise to the material isomorphism $\underset{\sim}{I}$ which has the property that $\underset{\sim}{E}_p \equiv \underset{\sim}{E}_q \circ \underset{\sim}{I}$ identically on P_p.

As S_p^* is independent of p, when it is computed relative to the reference configurations $\underset{\sim}{\zeta}_{\alpha,p}$ and $\underset{\sim}{\eta}_{\alpha,p}$ we can define the response function $\underset{\sim}{S}_\alpha^*$ relative to the chart $(U_\alpha, \underset{\sim}{\zeta}_\alpha, \underset{\sim}{\eta}_\alpha)$ by $S_\alpha^* \equiv S_p^*$, $p \in U_\alpha$; the domain of S_α^* is $P_\alpha \equiv P_p$, $\forall p \in U_\alpha$. Following Wang [32] we now denote the relative phase isotropy group of the chart and the relative index isotropy group of the chart by G_α and I_α, respectively, where

$$
\begin{aligned}
G_\alpha &\equiv \underset{\sim}{\zeta}_{\alpha,p} \circ G(P_p) \circ \underset{\sim}{\zeta}_{\alpha,p}^{-1} \\
I_\alpha &\equiv \underset{\sim}{\eta}_{\alpha,p} \circ I(P_p) \circ \underset{\sim}{\eta}_{\alpha,p}^{-1}
\end{aligned}
\qquad \text{(III-14)}
$$

for all $p \in U_\alpha$.

Remark Note that $G_\alpha \equiv G_p$ and $I_\alpha \equiv I_p$ for all $p \in U_\alpha$.

The following definition of the compatibility of two material-index charts in $\mathcal{B}$ is the natural generalization of the definition of compatibility which was introduced in Chapter I for reference charts on simple elastic bodies, i.e., we make

Definition III-12 Two material-index charts in $\mathcal{B}$, say, $(\mathcal{U}_\alpha, \underset{\sim}{\zeta}_\alpha, \underset{\sim}{\eta}_\alpha)$ and $(\mathcal{U}_\beta, \underset{\sim}{\zeta}_\beta, \underset{\sim}{\eta}_\beta)$, with $\mathcal{U}_\alpha \cap \mathcal{U}_\beta \neq \emptyset$, are called compatible if $\hat{G}_{\alpha\beta}(p) \equiv \underset{\sim}{\zeta}_{\beta,p}^{-1} \circ \underset{\sim}{\zeta}_{\alpha,p} \in G(P_p)$ and $\hat{I}_{\alpha\beta}(p) \equiv \underset{\sim}{\eta}_{\beta,p}^{-1} \circ \underset{\sim}{\eta}_{\alpha,p} \in I(P_p)$ for each $p \in \mathcal{U}_\alpha \cap \mathcal{U}_\beta$.

It follows at once from this last definition that $\underset{\sim}{S}_\alpha^* = \underset{\sim}{S}_\beta^*$, $G_\alpha = G_\beta$ and $I_\alpha = I_\beta$; however, unlike the situation which prevailed for the case of simple elastic particles (i.e., see the remark on page II-15) the converse of this statement is false. By analogy with the presentation for simple elastic particles, we call the fields defined via $G_{\alpha\beta}(p) \equiv \underset{\sim}{\zeta}_{\beta,p} \circ \underset{\sim}{\zeta}_{\alpha,p}^{-1}$ and $I_{\alpha\beta}(p) \equiv \underset{\sim}{\eta}_{\beta,p} \circ \underset{\sim}{\eta}_{\alpha,p}^{-1}$, $\forall p \in \mathcal{U}_\alpha \cap \mathcal{U}_\beta$, the coordinate transformations from the material-index chart $(\mathcal{U}_\alpha, \underset{\sim}{\zeta}_\alpha, \underset{\sim}{\eta}_\alpha)$ to the material-index chart $(\mathcal{U}_\beta, \underset{\sim}{\zeta}_\beta, \underset{\sim}{\eta}_\beta)$ and by virtue of the definition of compatibility given above, and (III-4), we see that $G_{\alpha\beta}(p) \in G_\alpha \equiv G_\beta$ and $I_{\alpha\beta}(p) \in I_\alpha \equiv I_\beta$ for all $p \in \mathcal{U}_\alpha \cap \mathcal{U}_\beta$. Also, because $\underset{\sim}{\zeta}_\alpha(\underset{\sim}{\zeta}_\beta)$ and $\underset{\sim}{\eta}_\alpha(\underset{\sim}{\eta}_\beta)$ are diffeomorphisms the coordinate transformations $G_{\alpha\beta}$ and $I_{\alpha\beta}$ are smooth fields on $\mathcal{U}_\alpha \cap \mathcal{U}_\beta$

We are now in a position to define a material-index atlas for $\mathcal{B}$ which we take to be a collection

$\underset{\sim}{T} = \{(U_\alpha, \underset{\sim}{\zeta}_\alpha, \underset{\sim}{\eta}_\alpha), \alpha\varepsilon I\}$ of mutally compatible material-index charts in $\mathcal{B}$, where $\{U_\alpha, \alpha\varepsilon I\}$ is an open covering of $\mathcal{B}$; we also require that T be maximal with respect to these two conditions. If $\mathcal{B}$ can be equipped with a material-index atlas then we will call $\mathcal{B}$ a smooth materially uniform generalized elastic body.

Now, as all the material-index charts in a material-index atlas are mutually compatible, it should be clear that the relative response function $\underset{\sim}{S}^*_\alpha$ with respect $(U_\alpha, \underset{\sim}{\zeta}_\alpha, \underset{\sim}{\eta}_\alpha) \in \underset{\sim}{T}$, is independent of α. We can, therefore, define a response function $\underset{\sim}{S}^*_{\underset{\sim}{T}}$ with respect to the atlas $\underset{\sim}{T}$ by $\underset{\sim}{S}^*_{\underset{\sim}{T}} \equiv \underset{\sim}{S}^*_\alpha$ for all $\alpha\varepsilon I$; the domain of $\underset{\sim}{S}^*_{\underset{\sim}{T}}$ is denoted by $P_{\underset{\sim}{T}} \equiv P_\alpha$, $\forall \alpha\varepsilon I$. In a similar fashion the relative phase isotropy group $G_{\underset{\sim}{T}}$ and the relative index isotropy group $I_{\underset{\sim}{T}}$ are defined by $G_{\underset{\sim}{T}} = G_\alpha$, $\forall\alpha\varepsilon I$ and $I_{\underset{\sim}{T}} = I_\alpha$, $\forall\alpha\varepsilon I$, respectively; they are, of course, the phase and index isotropy groups for the response function $\underset{\sim}{S}^*_{\underset{\sim}{T}}$

Remark It is important to note (i.e., Wang [32], §2) that it is possible to have two material-index atlases $\underset{\sim}{T}$ and $\bar{\underset{\sim}{T}}$ with $\underset{\sim}{T} \neq \bar{\underset{\sim}{T}}$ and yet $\underset{\sim}{S}_{\underset{\sim}{T}} \equiv \underset{\sim}{S}_{\bar{\underset{\sim}{T}}}$. In fact suppose that $\underset{\sim}{K}: R^3 \to R^3$ and $\underset{\sim}{\xi}: V \to V$ are diffeomorphisms with $\underset{\sim}{K} \notin G_{\underset{\sim}{T}}$ $\underset{\sim}{\xi} \notin I_{\underset{\sim}{T}}$ but $(\underset{\sim}{F}\underset{\sim}{K}, \underset{\sim}{\xi}(y)) \in P_{\underset{\sim}{T}}$ and $\underset{\sim}{S}^*_{\underset{\sim}{T}}(\underset{\sim}{F}\underset{\sim}{K}, \underset{\sim}{\xi}(y)) = \underset{\sim}{S}^*_{\underset{\sim}{T}}(\underset{\sim}{F}, y)$ for for all $(\underset{\sim}{F}, y) \in P_{\underset{\sim}{T}}$. Then $\underset{\sim}{T} \equiv \{(U_\alpha, \underset{\sim}{\zeta}_\alpha, \underset{\sim}{\eta}_\alpha), \alpha\varepsilon I\}$ and $\bar{\underset{\sim}{T}} \equiv \{(U_\alpha, \underset{\sim}{K}\circ\underset{\sim}{\zeta}_\alpha, \underset{\sim}{\xi}^{-1}\circ\underset{\sim}{\eta}_\alpha), \alpha\varepsilon I\}$ are distinct material-index atlases on $\mathcal{B}$ but, by virtue of the definition of $\underset{\sim}{S}^*_p$, for any $p \varepsilon \mathcal{U}_\alpha$ (i.e., (III-11)), it is clear that

$\underset{\sim}{S}^*_{\underset{\sim}{T}}(\underset{\sim}{F},y) = \underset{\sim}{S}^*_{\underset{\sim}{T}}(\underset{\sim}{F}\underset{\sim}{K},\xi(y)) = \underset{\sim}{S}^*_{\underset{\sim}{T}}(\underset{\sim}{F},y)$. Thus, we lose(6) the one-to-one correspondence between (material-index) atlases $\underset{\sim}{T}$ and their relative response functions $\underset{\sim}{S}^*_{\underset{\sim}{T}}$ which prevailed for the analogous atlases and response functions on simple elastic bodies.

<u>Exercise</u> 10 If $\underset{\sim}{L}: R^3 \to R^3$ and $\underset{\sim}{\lambda}: \mathcal{Y} \to \mathcal{Y}$ are arbitrary diffeomorphisms, show that $\hat{\underset{\sim}{T}} = \{(\mathcal{U}_\alpha, \underset{\sim}{L}\circ\underset{\sim}{\zeta}_\alpha, \underset{\sim}{\lambda}^{-1}\circ\underset{\sim}{\eta}_\alpha), \alpha\varepsilon I\}$ is a material-index atlas for $\mathcal{B}$ whenever $\underset{\sim}{T} = \{(\mathcal{U}_\alpha,\zeta_\alpha,\eta_\alpha), \alpha\varepsilon I\}$ is. Show also that the following transformation laws hold for the response functions, their domains, and their isotropy groups relative to $\underset{\sim}{T}$ and $\hat{\underset{\sim}{T}}$: if $(\underset{\sim}{F},y) \;\varepsilon\; \mathcal{P}_{\hat{\underset{\sim}{T}}}$ then $(\underset{\sim}{F}\underset{\sim}{L},\underset{\sim}{\lambda}(y)) \;\varepsilon\; \mathcal{P}_{\underset{\sim}{T}}$ and

$$\underset{\sim}{S}^*_{\hat{\underset{\sim}{T}}}(\underset{\sim}{F},y) \equiv \underset{\sim}{S}^*_{\underset{\sim}{T}}(\underset{\sim}{F}\underset{\sim}{L},\underset{\sim}{\lambda}(y)) \tag{III-15}$$

$$\mathcal{G}_{\hat{\underset{\sim}{T}}} = \underset{\sim}{L}\mathcal{G}_{\underset{\sim}{T}}\underset{\sim}{L}^{-1}$$

$$\mathcal{I}_{\hat{\underset{\sim}{T}}} = \underset{\sim}{\lambda}^{-1}\circ\mathcal{I}_{\underset{\sim}{T}}\circ\underset{\sim}{\lambda}$$

Finally, if $\underset{\sim}{K}$ and $\underset{\sim}{\xi}$ are diffeomorphisms of R^3 and $\mathcal{Y}$, respectively, which have the property delineated in the remark above, then

$$\underset{\sim}{K}\mathcal{G}_{\underset{\sim}{T}}\underset{\sim}{K}^{-1} = \mathcal{G}_{\underset{\sim}{T}} \tag{III-16}$$

$$\underset{\sim}{\xi}\circ\mathcal{I}_{\underset{\sim}{T}}\circ\underset{\sim}{\xi}^{-1} = \mathcal{I}_{\underset{\sim}{T}}$$

(6) this correspondence <u>can</u> be assumed valid when we derive the field equations of motion, as the exception noted in the remark above does not come into play.

6. Material Tangent Bundles and Index Bundles; the Material and Index Atlases, Homogeneity and Local Homogeneity.

In Chapter I it was necessary to deal only with the material tangent bundle $T(B,\underset{\sim}{\Phi})$, of a simple elastic body B, relative to a given reference atlas $\underset{\sim}{\Phi}$. If B is a generalized elastic body, however, and $\underset{\sim}{T} = \{(U_\alpha, \zeta_\alpha, \eta_\alpha), \alpha\varepsilon I\}$ is a material-index atlas for B then we can define both a material atlas $\zeta'(\underset{\sim}{T})$ and an index atlas $\eta'(\underset{\sim}{T})$ relative to $\underset{\sim}{T}$ via

$$\zeta'(\underset{\sim}{T}) \equiv \{(U_\alpha, \underset{\sim}{\zeta}_\alpha), \alpha\varepsilon I\} \qquad \text{(III-17)}$$

$$\eta'(\underset{\sim}{T}) \equiv \{(U_\alpha, \underset{\sim}{\eta}_\alpha), \alpha\varepsilon I\}$$

and by definition these sets now form the prebundle atlases for the material tangent bundle $T(B,\underset{\sim}{T})$ and the index bundle $V(B,\underset{\sim}{T})$, respectively, with respect to $\underset{\sim}{T}$. Recall now that $\{U_\alpha, \alpha\varepsilon I\}$ forms an open cover for B and that

(i′) for each $\alpha\varepsilon I$, $\underset{\sim}{\zeta}_\alpha$ and $\underset{\sim}{\eta}_\alpha$ are diffeomorphisms satisfying conditions (i) and (ii) of Definition III-11

(ii′) the coordinate transformations $G_{\alpha\beta}$ and $I_{\alpha\beta}$ (defined following Definition III-12) are smooth fields which take their values in $G_{\underset{\sim}{T}}$ and $I_{\underset{\sim}{T}}$, respectively.

We can, therefore, maximize the prebundle atlases $\zeta'(\underset{\sim}{T})$ and $\eta'(\underset{\sim}{T})$ with respect to conditions (i′) and (ii′) above and, in so doing, we are led to the following

Definition III-13 The unique maximal sets

$$\zeta(\underset{\sim}{T}) = \{(U_\alpha, \underset{\sim}{\zeta}_\alpha),\ \alpha\varepsilon J\}$$

$$\eta(\underset{\sim}{T}) = \{(U_\alpha, \underset{\sim}{\eta}_\alpha),\ \alpha\varepsilon K\}$$

obtained by maximizing $\zeta'(\underset{\sim}{T})$ and $\eta'(\underset{\sim}{T})$, with respect to conditions (i′) and (ii′) above are the bundle atlases for $T(B,\underset{\sim}{T})$ and $V(B,\underset{\sim}{T})$, respectively. The fibre space and structure group of $T(B,\underset{\sim}{T})$ are R^3 and $G_{\underset{\sim}{T}}$ while those of $V(B,\underset{\sim}{T})$ are V and $I_{\underset{\sim}{T}}$.

Exercise 11 If $\hat{\zeta}'(\underset{\sim}{T})$ and $\hat{\eta}'(\underset{\sim}{T})$ are, respectively, the material atlas and the index atlas relative to the material-index atlas $\underset{\sim}{T}$, which is defined in the statement of exercise 10, show that the bundle atlases $\hat{\zeta}(\hat{\underset{\sim}{T}})$ and $\hat{\eta}(\hat{\underset{\sim}{T}})$ of $T(B,\hat{\underset{\sim}{T}})$ and $V(B,\hat{\underset{\sim}{T}})$, respectively, are related to those of $T(B,\underset{\sim}{T})$ and $V(B,\underset{\sim}{T})$ via

$$\hat{\zeta}(\hat{\underset{\sim}{T}}) = \{(U_\alpha,\ \underset{\sim}{L}\circ\underset{\sim}{\zeta}_\alpha),\ \alpha\varepsilon J\}$$
$$\hat{\eta}(\hat{\underset{\sim}{T}}) = \{(U_\alpha,\ \underset{\sim}{\lambda}^{-1}\circ\underset{\sim}{\eta}_\alpha),\ \alpha\varepsilon K\} \qquad \text{(III-18)}$$

As was the case for simple elastic bodies, we can define, in an analogous manner, the notions of homogeneous and locally homogeneous generalized elastic bodies, i.e., we say that B is homogeneous if there exists a global material-index chart, say $(B,\underset{\sim}{\zeta},\underset{\sim}{\eta})$, and a configuration $\phi: B \to R^3$ such that $\underset{\sim}{\zeta}_p = \phi_{*p}$, $\forall p\varepsilon B$. Similarly, B is said to be locally homogeneous if there exists a (local) material index chart, say $(U_\alpha, \underset{\sim}{\zeta}_\alpha, \underset{\sim}{\eta}_\alpha)$ and a configuration $\phi_\alpha: U_\alpha \to R^3$ such that $p \varepsilon U_\alpha$ and $\underset{\sim}{\zeta}_{\alpha p} = \phi_{\alpha * p}$, $\forall p \varepsilon U_\alpha$. We can also make

Definition III-14. The material tangent bundle $T(B, \underset{\sim}{T})$ is homogeneous if the bundle atlas $\zeta(\underset{\sim}{T})$ has a global chart $(B, \underset{\sim}{\zeta})$ such that $\underset{\sim}{\zeta}_p = \phi_{*p}$ for some configuration $\phi: B \to R^3$ and all $p \varepsilon B$; similarly, $T(B, \underset{\sim}{T})$ is locally homogeneous if for each $p \varepsilon B$ there exists a chart $(U_\alpha, \underset{\sim}{\zeta}_\alpha) \in \zeta(\underset{\sim}{T})$ such that $p \varepsilon U_\alpha$ and $\underset{\sim}{\zeta}_{\alpha p} = \phi_{\alpha * p}$ for all $p \varepsilon U_\alpha$ and some configuration $\phi_\alpha: U_\alpha \to R^3$. Also, we say that $V(B, \underset{\sim}{T})$ is homogeneous if there exists a global chart $(B, \underset{\sim}{\eta})$ in $\eta(\underset{\sim}{T})$.(7)

Exercise 12 Show that the conditions for the homogeneity and local homogeneity of $T(B, \underset{\sim}{T})$ are independent of the choice of the material-index atlas $\underset{\sim}{T}$.

Exercise 13 (Wang [32]; §3). Prove that B is homogeneous (or locally homogeneous) iff both $T(B, \underset{\sim}{T})$ and $V(B, \underset{\sim}{T})$ are homogeneous (or locally homogeneous). Can it be true that $T(B, \underset{\sim}{T})$ is homogeneous without the some being true for the index bundle $V(B, \underset{\sim}{T})$ relative to $\underset{\sim}{T}$?

(7) every fibre bundle is (locally) trivial so $V(B, \underset{\sim}{T})$ is, a priori, locally homogeneous.

7. Material and Index Connections

In Chapter II we introduced the concept of a material connection on a simple elastic body $\mathcal{B}$; this was defined as being a "G" connection on $E(\mathcal{B},\underset{\sim}{\Phi})$ where $\underset{\sim}{\Phi}$ is some reference atlas for $\mathcal{B}$. In equivalent terms, a material connection H on $\mathcal{B}$, a simple elastic body, is an affine connection of $T(\mathcal{B},\underset{\sim}{\Phi})$ whose induced parallel transports of the tangent spaces on $\mathcal{B}$ are material isomorphisms. The parallel transports $\rho_t\colon E_{\lambda(0)} \to E_{\lambda(t)}$, where λ is a smooth curve $\mathcal{B}$, may be extended to well defined maps $\hat{\rho}_t\colon \mathcal{B}_{\lambda(0)} \to \mathcal{B}_{\lambda(t)}$ and themselves induce maps $\rho_{t,\alpha}\colon R^3 \to R^3$ which are defined via (II-14) with $\lambda \subset U_\alpha$ and $(U_\alpha,\xi_\alpha) \in \xi(\underset{\sim}{\Phi})$. (Recall that $\xi_\alpha\colon U_\alpha \times G(\underset{\sim}{\Phi}) \to \pi^{-1}(U_\alpha)$ and that the fibre space of $E(\mathcal{B},\underset{\sim}{\Phi})$, i.e., R^3 coincides with the structure group $G(\underset{\sim}{\Phi})$) The condition that the maps $\hat{\rho}_t$ be material isomorphisms, in the sense of Chapter II, is then equivalent to the requirement that the transformations $\rho_{t,\alpha}$ form smooth curves in $G(\underset{\sim}{\Phi})$ which pass through the identity element at $t = 0$; indeed, this is the condition which led to the stated restriction, in Theorem II-3, on the connection symbols Γ^i_{jk} of H and it is this restriction, given in terms of the Lie algebra $\underset{\sim}{g}(\underset{\sim}{\Phi})$ associated with $G(\underset{\sim}{\Phi})$, which characterizes material connections on simple elastic bodies.

Now, let $\underset{\sim}{T}$ be a material-index atlas for $\mathcal{B}$, a smooth materially uniform generalized elastic body. Because the structure on the material tangent bundle $T(\mathcal{B},\underset{\sim}{T})$ is essentially the same as that on $T(\tilde{B},\underset{\sim}{\Phi})$, with $\tilde{B}$ a simple elastic body and $\underset{\sim}{\Phi}$ a reference atlas on $\tilde{B}$, it is easy to see that if we define a <u>material connection on $T(B)$</u> to be a "G" connection on $T(B,\underset{\sim}{T})$ a condition formally equivalent to that stated in Theorem II-3 is applicable here. In other words, if (x^i) is a fixed global coordinate system on B induced by $\phi\colon B \to R^3$ then any chart $(U_\alpha, \underset{\sim}{\zeta}_\alpha)$ in $\zeta(\underset{\sim}{T})$ can be characterized by

$$\underset{\sim}{\zeta}_\alpha = \zeta^i_{\cdot j}\, dx^i \otimes \underset{\sim}{e}_i$$

and the parallel transport from a reference point $\lambda(0)$ to any arbitrary point $\lambda(t)$ on $\lambda \subset U_\alpha$ has the component form

$$\sigma_t = \sigma^i_{\cdot j}(t) d_{\lambda(0)} x^i \otimes \left.\frac{\partial}{\partial x^i}\right|_{\lambda(t)}$$

where $\sigma^i_{\cdot j}(0) = \delta^i_{\cdot j}$ and

$$\dot{\sigma}^i_{\cdot j} + \Gamma^i_{lk}\, \sigma^l_{\cdot j} \dot{\lambda}^k = 0, \quad i,j = 1,2,3$$

The map $\sigma_t(\alpha) \equiv \underset{\sim}{\zeta}_{\alpha,\lambda(t)} \circ \sigma_t \circ \zeta^{-1}_{\alpha,\lambda(0)}$, taken relative to the chart $(U_\alpha, \underset{\sim}{\zeta}_\alpha)$, is an isomorphism of R^3 which can be written in component form as

$$\sigma_t(\alpha) = \xi^i_{\cdot j}(t) \underset{\sim}{e}_i \otimes \underset{\sim}{e}^j$$

In order for H, with connection symbols Γ^i_{jk} relative to (x^i), to be a "G" connection on $T(B,\underset{\sim}{T})$, we require that $\sigma_t(\alpha)$ be an integral curve of the Lie algebra $\underset{\sim}{g}(\underset{\sim}{T})$ of G_T; this then leads to the condition that the matrices

$$\{(\zeta^{-1})^i_j \left[\frac{\partial \zeta^i_1}{\partial x^m} - \Gamma^k_{1m}\zeta^i_k\right], \; m = 1,2,3\}$$

belong to $\underset{\sim}{g}(\underset{\sim}{T})$ at each $p\varepsilon U_\alpha$.

Thus far, everything is pretty much the same as it was in Chapter II; however, as we pass from simple elastic bodies $\tilde{B}$ to generalized elastic bodies B, we introduce into the general picture the index bundle $V(B,\underset{\sim}{T})$ relative to a material-index atlas $\underset{\sim}{T}$. So, following Wang ([32], §3), we can now define an <u>index connection</u> V on $V(B)$ as a "G" connection on $V(B,\underset{\sim}{T})$, for any material-index atlas $\underset{\sim}{T}$.

In order to deduce the conditions which characterize an index connection on $V(B)$ we must consider the Lie algebra $\underset{\sim}{i}_{\underset{\sim}{T}}$ associated with $\underset{\sim}{I}_{\underset{\sim}{T}}$ and the Lie algebra $\underset{\sim}{y}_{\underset{\sim}{T}}$ of Y which is induced by $\underset{\sim}{I}_{\underset{\sim}{T}}$. Thus, let $u\varepsilon\underset{\sim}{i}_{\underset{\sim}{T}}$ and consider the one-parameter Lie transformation group on Y which is given by

$$\phi_{\underset{\sim}{u}}(s) = \exp(s\underset{\sim}{u}), \quad s\varepsilon R.$$

This Lie transformation group can be characterized by its <u>infinitesmal generator</u> $\bar{\underset{\sim}{u}}$, i.e., by the vector field defined via

$$\bar{\underset{\sim}{u}}(y) \equiv \frac{d}{ds}\,[\phi_{\underset{\sim}{u}}(s)](y)\Big|_{s=0}, \quad \forall y\varepsilon Y \qquad \text{(III-19)}$$

We note that (III-19) defines a map of $\underset{\sim}{i}_{\underset{\sim}{T}}$ onto a set of vector fields on Y; we denote this set by $\underset{\sim}{y}_{\underset{\sim}{T}}$ and we call it the <u>Lie algebra of Y</u> (induced by $\underset{\sim}{I}_{\underset{\sim}{T}}$). The bracket operation on this Lie algebra is just the usual one induced by the Lie derivative, i.e., if $\bar{\underset{\sim}{u}},\bar{\underset{\sim}{v}} \in \underset{\sim}{\mathcal{Y}}_T$ then $[\bar{\underset{\sim}{u}},\bar{\underset{\sim}{v}}] = L_{\bar{\underset{\sim}{u}}}\bar{\underset{\sim}{v}}$

Once again we take (x^i) as a global coordinate system on $\mathcal{B}$ and let $(y^\delta,\ \delta=1,2,..n)$ be a local coordinate system in $\mathcal{Y}$. Also, let $(\mathcal{U}_\alpha,\underset{\sim}{\eta}_\alpha) \in \eta(\underset{\sim}{T})$ be an index chart. Then, as $\underset{\sim}{\eta}_\alpha\colon \mathcal{U}_\alpha \times \mathcal{Y} \to \mathcal{V}(\mathcal{U}_\alpha)$, this diffeomorphism induces a map of the local coordinate system $(x^i,y^\delta) \to (x^i,y^\delta)_\alpha$, which is a local coordinate system in $\mathcal{V}(\mathcal{U}_\alpha)$. In other words, for each $p\in\mathcal{U}_\alpha$, $(x^i(p),\ y^\delta)_\alpha$ is a local coordinate system in $\mathcal{V}_p$, the fibre over p. Now let $\pi\colon \mathcal{V}(\mathcal{B}) \to \mathcal{B}$ denote the projection map so that $\pi^{-1}(p) = \mathcal{V}_p$, $p\in\mathcal{B}$. Then, $\pi_*(\frac{\partial}{\partial y^\delta})_\alpha = 0$, $\delta=1,2,..$ so that the basis vectors $\{\frac{\partial}{\partial y^\delta}\}_\alpha$ are vertical vectors in $\mathcal{V}(\mathcal{B})$. In terms of a connection V on $\mathcal{V}(\mathcal{B})$ and the local coordinate system $(x^i,y^\delta)_\alpha$, horizontal subspaces relative to V are spanned by sets of the form

$$\{\frac{\partial}{\partial x^i} - \mathcal{V}_i^\delta(x,y)\frac{\partial}{\partial y^\delta},\ i = 1,2,3\}$$

and we call the tensor field

$$\overset{\alpha}{\underset{\sim}{\mathcal{V}}} \equiv \mathcal{V}_j^\delta\ dx^j \otimes \frac{\partial}{\partial y^\delta}$$

the <u>connection</u> <u>form</u> of V with respect to the index chart $(\mathcal{U}_\alpha,\underset{\sim}{\eta}_\alpha)$. If $\lambda \subset \mathcal{U}_\alpha$ is a smooth curve then the induced parallel transport $\gamma_t(\alpha)$ along λ with respect to V is determined by solving the equations of parallel transport which, in this case, assume the form

$$\dot{y}^\delta + V_i^\delta(x,y)\dot{\lambda}^i = 0, \ \delta = 1,2,\dots, n. \qquad \text{(III-20)}$$

In other words, $[\gamma_t(\alpha)]\ (y(0) = y(t)$, where $y(t)$ is any solution of (III-20). So, just as we required that the transformations $\sigma_t(\alpha)$ of R^3, the fibre space of $T(B,\underset{\sim}{T})$, form a smooth curve in $G_{\underset{\sim}{T}}$, we now require that the transformations $\gamma_t(\alpha)$ of V, the fibre space of $V(B,\underset{\sim}{T})$, form a smooth curve in $I_{\underset{\sim}{T}}$; this is, however, equivalent to requiring that the infinitesmal generator of $\gamma_t(\alpha)$ be a member of the Lie algebra $\underset{\sim}{y}_{\underset{\sim}{T}}$ of V, which is induced by $I_{\underset{\sim}{T}}$, <u>for each t</u>. A simple computation employing the equations of parallel transport III-20 yields

$$\begin{aligned}\frac{d}{dt}[\gamma_t(\alpha)](y(0)) &= \dot{y}(t)\\ &= \dot{y}^\delta(t)\frac{\partial}{\partial y^\delta}\bigg|_{y(t)}\\ &= -V_i^\delta(\lambda(t),y(t))\dot{\lambda}^i(t)\frac{\partial}{\partial y^\delta}\bigg|_{y(t)}\end{aligned}$$

and leads us to the following

<u>Theorem</u> III-4. A connection V on $V(B)$ is an index connection (i.e., a "G" connection on $V(B,\underset{\sim}{T})$ relative to <u>any</u> material-atlas $\underset{\sim}{T}$) iff the components V_j^δ of the connection form of V with respect to the index chart $(U_\alpha, \underset{\sim}{\eta}_\alpha) \in \underset{\sim}{T}$ (and the local coordinate system (x^i, y^δ) on $U_\alpha \times V$) satisfy,

$$V_j^{\delta}(p,\cdot)\frac{\partial}{\partial y^{\delta}} \quad \varepsilon \ \underset{\sim}{\mathcal{Y}}_T, \quad \forall p \varepsilon \mathcal{U}_{\alpha;} \ j=1,2,3$$

Remark It is a simple matter to prove that the tensor field V_{α} (the connection form of V with respect to $(\mathcal{U}_{\alpha},\underset{\sim}{\eta}_{\alpha})$) is independent of the choice of coordinate systems (x^i,y^{δ}); the same sort of invariance does not hold, however, with regard to a change of the index chart.

8. Field Equations of Motion in Generalized Elasticity

Following Wang ([32], §4) we will now go through the derivation of the field equations of motion for generalized elastic bodies $\mathcal{B}$ in a single phase, say, $P(\mathcal{B})$. In order to obtain the equations for motion in a global form we proceed as we did in Chapter II making use, in the present situation, however, of the concepts of both material and index connections.

So, let $\mathcal{B}$ be a smooth materially uniform generalized elastic body; we take $\underset{\sim}{T}$ as our fixed material-index atlas for $\mathcal{B}$ in the fixed phase $P(\mathcal{B})$. Material-index charts in $\underset{\sim}{T}$ are of the form $(\mathcal{U}_{\alpha},\underset{\sim}{\zeta}_{\alpha},\underset{\sim}{\eta}_{\alpha})$, $\alpha\varepsilon I$, and the relative response function with respect to $\underset{\sim}{T}$ is $\overset{*}{\underset{\sim}{S}}_{\underset{\sim}{T}}$. As there is no possibility here of confusion between $\overset{*}{\underset{\sim}{S}}_{\underset{\sim}{T}}$ and the relative response function $\underset{\sim}{S}_{\underset{\sim}{\Phi}}$ of Chapter II, we shall, for the sake of simplicity in writing, drop both the asterisk and the subscript $\underset{\sim}{T}$ on the response function $\overset{*}{\underset{\sim}{S}}_{\underset{\sim}{T}}$; the constitutive equation for the stress tensor $\underset{\sim}{T}_S$ then has the form

$$\underset{\sim}{T}_S = \underset{\sim}{S}(\underset{\sim}{F},y) \tag{III-23}$$

and relative to the standard basis for $\mathcal{R}^3$ and a local coordinate system (y^δ) in $\mathcal{Y}$ we have the component form

$$T^{ij} = S^{ij}(\underset{\sim}{F},y) \tag{III-24}$$

where we have dropped the subscript on the stress tensor. Of great importance to us in what follows are several results which are obtained in a manner analogous to the way in which we derived (II-27) for simple elastic bodies. So, under the usual smoothness assumptions on the functions S^{ij}, we begin by defining the functions S^{ij}_{kl} and S^{ij}_{δ} as the components of the gradient of $\underset{\sim}{S}$ with respect to the natural bases in $\mathcal{R}^3$ and $\mathcal{Y}$, i.e.

$$S^{ij}_{kl} = S^{ij}_{kl}(\underset{\sim}{F},y) \equiv \partial S^{ij}/\partial F^{kl} \tag{III-25}$$

$$S^{ij} = S^{ij}_{\delta}(\underset{\sim}{F},y) \equiv \partial S^{ij}/\partial y^{\delta}$$

Recall now that

$$\underset{\sim}{S}(\underset{\sim}{F}\underset{\sim}{K},y) = \underset{\sim}{S}(\underset{\sim}{F},y) \tag{III-26}$$

$$\underset{\sim}{S}(\underset{\sim}{F},\xi(y)) = \underset{\sim}{S}(\underset{\sim}{F},y)$$

for all $\underset{\sim}{K}\in G_T$ and all $\xi\in\underset{\sim}{I}_T$. If we take the gradient on both sides of (III-26_1), with respect to the natural basis in R^3, we easily obtain

$$S^{ij}_{ks}(\underset{\sim}{F}\underset{\sim}{K},y)K^s_l = S^{ij}_{kl}(\underset{\sim}{F},y) \qquad \text{(III-27)}$$

On the other hand, if we take the gradient on both sides of (III-26_2), with respect to the natural basis in Y, we get

$$S^{ij}_{\delta}(\underset{\sim}{F},\xi(y))\frac{\partial\xi^{\delta}}{\partial y^{\nu}}(y) = S^{ij}_{\nu}(\underset{\sim}{F},y) \qquad \text{(III-28)}$$

Now let $\underset{\sim}{K} = \underset{\sim}{K}(t)$ and $\xi = \xi(t)$ be smooth curves in $G_{\underset{\sim}{T}}$ and $I_{\underset{\sim}{T}}$ which pass through the respective identity elements of these isotropy groups at $t = 0$. Then differentiating equation (III-26_1) through with respect to t yields (after setting $t = 0$):

$$S^{ij}_{kl}(\underset{\sim}{F},y)F^k_s D^{sl} = 0 \qquad \text{(III-29)}$$

where $\underset{\sim}{D} = \dot{\underset{\sim}{K}}(0) \in \underset{\sim}{g}(\underset{\sim}{T})$, the Lie algebra of the isotropy group $G_{\underset{\sim}{T}}$. Finally, differentiating equation (III-26_2) with respect to t (and then setting $t = 0$) yields

$$S^{ij}_{\delta}(\underset{\sim}{F},y)\bar{\underset{\sim}{u}}^{\delta}(y) = 0 \qquad \text{(III-30)}$$

where $\frac{d}{dt}[\xi(t)y]\Big|_{t=0} \equiv \bar{\underset{\sim}{u}}(y) \in \underset{\sim}{Y}_{\underset{\sim}{T}}$, the Lie algebra of Y, which is induced by $I_{\underset{\sim}{T}}$. (Note that $\xi(t)y \subset Y$ is a smooth curve, for each $y \in Y$, which passes through the element

$y \in \mathcal{Y}$ at time $t = 0$). Obviously, (III-29) and (III-30) are valid for all $\underset{\sim}{D} \in \underset{\sim}{g}(\underset{\sim}{T})$ and $\bar{\underset{\sim}{u}} \in y_{\underset{\sim}{T}}$, respectively.

Now, let ϕ be a configuration of $\mathcal{B}$ which induces the global coordinate system (x^i), let θ be an index section, and suppose that $(\mathcal{U}_\alpha, \underset{\sim}{\zeta}_\alpha, \underset{\sim}{\eta}_\alpha)$ is an arbitrary material-index chart in $\underset{\sim}{T}$. If $x \in \mathcal{U}_\alpha$, then the stress tensor at $\phi(x)$ is determined via (III-23) with $\underset{\sim}{F} \equiv \phi_{*x} \circ \underset{\sim}{\zeta}_\alpha^{-1}(x)$ and $y \equiv \eta_\alpha^{-1}(\theta(x))$, i.e.

$$\underset{\sim}{T}(\phi(x)) = \underset{\sim}{S}(\phi_{*x} \circ \underset{\sim}{\zeta}_\alpha^{-1}(x), \underset{\sim}{\eta}_\alpha(\theta(x))$$

or, in component form

$$T^{ij}(\phi(x)) = S^{ij}((\underset{\sim}{\zeta}^{-1})^k_l, \theta^\delta) \qquad \text{(III-31)}$$

where the α subscript has been dropped. As in Chapter II, the Cauchy equation of motion is given by (II-24) so subsitution of (III-31) into this equation yields

$$S^{ij}_{kl} \frac{\partial(\zeta^{-1})^{kl}}{\partial x^j} + S^{ij}_\delta \frac{\partial \theta^\delta}{\partial x^j} + \rho b^i = \rho \ddot{x}^i \qquad \text{(III-32)}$$

where the arguments of the gradient functions are the same as those of the S^{ij} in (III-31). As was the case in Chapter II, when we deduced the equation of motion (II-25), the above equation is a local one valid for points in the coordinate neighborhood $\mathcal{U}_\alpha$ only. To obtain an equation which is valid everywhere on $\mathcal{B}$, we will follow the lead of the derivation in Chapter II and make use of the conditions which characterize, respectively, material connections on $\mathcal{T}(\mathcal{B})$ and index connections on $\mathcal{V}(\mathcal{B})$. So, let H be a fixed material connection on $\mathcal{T}(\mathcal{B})$ and V a fixed index connection on $\mathcal{V}(\mathcal{B})$; as in §7, we represent H and V by the connection symbols Γ^i_{jk} and $\mathcal{V}^\delta_i$, respectively, relative to the coordinate systems (x^i) and (y^δ). Since (III-29) must hold for each $\underset{\sim}{D} \varepsilon \underset{\sim}{g}(\underset{\sim}{T})$ and the matrices

$$\{(\zeta^{-1})^l_j \, [\frac{\partial \zeta^i_j}{\partial x^m} - \Gamma^k_{lm} \, \zeta^i_k], \; m = 1,2,3\}$$

must belong to $\underset{\sim}{g}(\underset{\sim}{T})$ at each point $p\varepsilon\mathcal{U}_\alpha$, if the Γ^i_{jk} are to be the connection symbols of a material connection on $\mathcal{T}(\mathcal{B})$, we should obtain from (III-29) the equation

$$S^{ij}_{kl} \, [\frac{\partial (\zeta^{-1})^{kl}}{\partial x^j} + \Gamma^k_{sm} \, (\zeta^{-1})^{sl}] = 0 \qquad \text{(III-33)}$$

(Note that $\underset{\sim}{F} \equiv \phi_{*x} \circ \zeta^{-1}(x)$ and that

$$\frac{\partial \zeta^i_l}{\partial x^m} \, (\zeta^{-1})^l_j = - \zeta^i_k \, \frac{\partial (\zeta^{-1})^l_j}{\partial x^m}).$$

In a similar vein, we note that (III-30) must be valid for all $\underset{\sim}{\bar{u}}\varepsilon\mathcal{V}_{\underset{\sim}{T}}$ and (by virtue of (III-22) $\nu^\delta_j(p,\cdot)\frac{\partial}{\partial y^\delta}$ must belong to $\underset{\sim}{y}_{\underset{\sim}{T}}$, at each $p\varepsilon\mathcal{U}_\alpha$, if the ν^δ_j are to be the components of the connection form of an index connection on $\mathcal{V}(\mathcal{B})$. Therefore, combining (III-30) and (III-22) we obtain

$$S^{ij}_{\delta} \nu^{\delta}_{k} = 0 \qquad \text{(III-34)}$$

Before continuing the present line of argument we will pause to review certain ideas pertinent to the definition of "covariant derivative." In Chapter I (§6) we defined the operation of covariant differentiation by working with the linear isomorphisms of tensor spaces $\mathcal{T}^{r,s}_{\lambda}(t)$, over $\lambda \subset \mathcal{B}$, which are induced by the parallel transports along λ relative to a given connection. An alternative, but completely equivalent approach is the following (Wang, [32], §3):
Let θ: $\mathcal{B} \to \mathcal{V}(\mathcal{B})$ be the fixed index section which appears in (III-31). Then we may define the covariant derivative of θ with respect to V as the slope of θ at any point $(p, \theta(p)) \in \mathcal{V}(\mathcal{B})$ relative to the horizontal subspace (relative to V) at $(p, \theta(p))$. In other words, if $D_p\theta$ denotes the covariant derivative of θ at $p \in \mathcal{B}$ and $\underset{\sim}{v} \in \mathcal{B}_p$ is an arbitrary vector, then $[D_p\theta](\underset{\sim}{v})$ is the vertical component of $\theta_{*p}(\underset{\sim}{v}) \in \mathcal{V}(\mathcal{B})_{(p,\theta(p))}$. In order to obtain a component form for $D\theta$ we again take $(x^i, y^\delta)_\alpha$ as the local coordinate system in $\mathcal{V}(\mathcal{B})$ which is induced by the chart $(\mathcal{U}_\alpha, \underset{\sim}{\eta}_\alpha) \in \underset{\sim}{T}$ and the coordinate systems (x^i) in $\mathcal{B}$ and (y^δ) in $\mathcal{Y}$. We recall (§7 of this chapter) that the horizontal subspace (relative to V) at $(p, \theta(p))$ is spanned by the set

$$\left\{\frac{\partial}{\partial x^i}\bigg|_{(p,\theta(p))} - \mathcal{V}^{\delta}_{i}(p,\theta(p))\frac{\partial}{\partial y^\delta}\bigg|_{(p,\theta(p))},\ i = 1,2,3\right\} \qquad \text{(III-35)}$$

Let $\underset{\sim}{v} \varepsilon \mathcal{B}_p$ have the component form $\underset{\sim}{v} = v^i \left.\frac{\partial}{\partial x^i}\right|_p$ and $\theta\colon \mathcal{B} \to \mathcal{V}(\mathcal{B})$ the components $(\theta^\delta)_\alpha$. Then for $\theta_{*p}(\underset{\sim}{v}) \varepsilon \mathcal{V}(\mathcal{B})_{(p,\theta(p))}$ we compute

$$\theta_{*p}(\underset{\sim}{v}) = v^i \left.\frac{\partial}{\partial x^i}\right|_{(p,\theta(p))} + v^i \left.\frac{\partial \theta^\delta}{\partial x^i}\right|_p \left.\frac{\partial}{\partial y^\delta}\right|_{(p,\theta(p)} \tag{III-36}$$

and, using (III-35) we may break this vector in the tangent space to $\mathcal{V}(\mathcal{B})$ at $(p,\theta(p))$ up into its vertical and horizontal components, with the vertical component being given by

$$[D_p\theta](\underset{\sim}{v}) = v^i \left[\left.\frac{\partial \theta^\delta}{\partial x^i}\right|_p + \mathcal{V}^\delta_i(p,\theta(p))\right] \left.\frac{\partial}{\partial y^\delta}\right|_{(p,\theta(p))} \tag{III-37}$$

From (III-37) we now easily deduce ($\underset{\sim}{v} \varepsilon \mathcal{B}_p$ is arbitrary) that

$$D\theta = (D\theta)^\delta_i dx^i \otimes \frac{\partial}{\partial y^\delta} \tag{III-38}$$

where $(D\theta)^\delta_i \equiv \theta^\delta_{\ |i} = \frac{\partial \theta^\delta}{\partial x^i} + \mathcal{V}^\delta_i(\cdot,\ \theta(\cdot))$. Note that $\theta^\delta_{\ |i} = \mathcal{V}^\delta_i(\cdot,\theta)$ if the $(\theta^\delta)_\alpha$ are constants with respect to $(x^i,y^\delta)_\alpha$.

<u>Exercise</u> 14 Verify that the vertical component of $\theta_{*p}(\underset{\sim}{v}) \varepsilon \mathcal{V}(\mathcal{B})_{(p,\theta(p))}$ is given by (III-37).

To continue with our derivation of the global form of the field equations of motion for generalized elastic bodies we now be substitute for the expressions $S^{ij}_{kl} \frac{\partial (\zeta^{-1})^{kl}}{\partial x^j}$ and $S^{ij}_\delta \frac{\partial \theta^\delta}{\partial x^j}$ in (III-32) by using (III-33) and (III-34), respectively, and the definition of the covariant derivative $\theta^{\delta}_{\cdot|j}$ given above; in so doing we obtain the following system of equations for the motion of points in $\mathcal{B}$:

$$-S^{ij}_{kl}\,\Gamma^{k}_{sj}(\underset{\sim}{\zeta}^{-1})^{sl} + S^{ij}_{\delta}\,\theta^{\delta}_{\cdot|j} + \rho b^{i} = \rho\ddot{x}^{i} \qquad \text{(III-39)}$$

Exercise 15 Verify that the second term on the left-hand side of (III-39) is independent of the choice of the index chart $(\mathcal{U}_{\alpha},\underset{\sim}{\eta}_{\alpha}) \in \underset{\sim}{T}$.

Now, just as in Chapter II, we have a situation in which the functions Γ^{i}_{jk} and $V^{\delta}_{\cdot j}$ will be time-dependent in a motion $\phi(t)$: $\mathcal{B} \rightarrow R^{3}$, i.e., $\underset{\sim}{\Gamma} = \underset{\sim}{\Gamma}(\underset{\sim}{x})$, for instance, and in a motion $\phi(t)$ of $\mathcal{B}$, $\underset{\sim}{x} = \underset{\sim}{x}(t)$. Suppose we choose a fixed reference configuration κ: $\mathcal{B} \Rightarrow R^{3}$, as we did in the last chapter, where the problem of rendering explicit the time-dependence of $\underset{\sim}{\Gamma}$ arose; then the motion is again characterized by the deformations $x^{i} = x^{i}(X^{A},t)$, where κ induces the coordinate system X^{A} on $\mathcal{B}$. The transformation from $\Gamma^{i}_{jk}(\underset{\sim}{x}) \Rightarrow \overset{\kappa}{\Gamma}{}^{A}_{BC}(X)$ as we go from $(x^{i}) \rightarrow (X^{A})$ is still given by the rule of §7, Chapter I, while the transformations from $V^{\delta}_{\cdot j}(x) \rightarrow \overset{\kappa}{V}{}^{\delta}_{A}(X)$ and $\zeta^{i}_{\cdot j}(\underset{\sim}{x}) \rightarrow \zeta^{i}_{A}(\underset{\sim}{X})$ are given via

$$V^{\delta}_{\cdot i} = \overset{\kappa}{V}{}^{\delta}_{A}\,\frac{\partial X^{A}}{\partial x^{i}}\,, \quad \zeta^{i}_{\cdot j} = \overset{\kappa}{\zeta}{}^{i}_{A}\,\frac{\partial X^{A}}{\partial x^{j}}$$

Substituting for the functions $\Gamma^{i}_{jk}(x)$, $V^{\delta}_{\cdot j}(x)$ and $\zeta^{i}_{\cdot j}(x)$

in (III-39) then yields

$$S^{ij}_{kl}(\overset{\kappa}{\underset{\sim}{\zeta}}{}^{-1})^{Al}\frac{\partial X^B}{\partial x^j}\left(\frac{\partial^2 x^k}{\partial X^A \partial X^B} - \overset{\kappa}{\Gamma}{}^C_{AB}\frac{\partial x^k}{\partial X^C}\right)$$

$$+ S^{ij}_{\delta}\frac{\partial X^B}{\partial x^j}\overset{\kappa}{\theta}{}^{\delta}{}_{|B} + \rho b^i = \rho \ddot{x}^i \qquad \text{(III-40)}$$

Remarks In (III-40), $\overset{\kappa}{\theta}{}^{\delta}{}_{|B} \equiv \frac{\partial \overset{\kappa}{\theta}}{\partial X^B} + \overset{\kappa}{V}{}^{\delta}_{A}$ still depends, in general, on both (X^A) and t as the index section θ may be changing during the motion; the governing equation for the time evolution of the index section has not been treated in [32].

Exercise 15 Show that the terms $S^{ij}_{kl}\,(\overset{\kappa}{\underset{\sim}{\zeta}}{}^{-1})^{Al}$ and $S^{ij}_{\delta}\,\overset{\kappa}{\theta}{}^{\delta}{}_{|B}$ in (III-40) are independent of the choice of the material-index chart $(U_\alpha, \underset{\sim}{\xi}_\alpha, \underset{\sim}{\eta}_\alpha)$ in $\underset{\sim}{T}$. Thus, if we define

$$\overset{\kappa}{S}{}^{ijA}_{k} \equiv S^{ij}_{kl}(\overset{\kappa}{\underset{\sim}{\zeta}}{}^{-1})^{Al}$$

$$\overset{\kappa}{S}{}^{ij}_{B} \equiv S^{ij}_{\delta}\,\overset{\kappa}{\theta}{}^{\delta}{}_{|B}$$

then the equations of motion (III-40) assume the global form

$$\overset{\kappa}{S}{}^{ijA}_{k}\left(\frac{\partial^2 x^k}{\partial X^A \partial X^B} - \overset{\kappa}{\Gamma}{}^C_{AB}\frac{\partial x^k}{\partial X^C}\right) + \overset{\kappa}{S}{}^{ij}_{B}\frac{\partial X^B}{\partial x^j} + \rho b^i = \rho \ddot{x}^i \qquad \text{(III-41)}$$

Remarks As in Chapter II we could present here the global form which the Cauchy equations of motion assume for a generalized elastic body if we employ the components of the Piola-Kirchoff stress tensor, i.e., $\overset{\kappa}{T}{}^{A}_{k}$, instead of the components $T^{i}_{\cdot j}$ of the Cauchy stress tensor (the relationship between $\underset{\sim}{T}_{S}$ and $\overset{\kappa}{\underset{\sim}{T}}_{S}$ is given on page II-43 and the equations of motion are again those of equation (II-32)). While it is certainly more convenient in some initial-boundary value problems to employ $\overset{\kappa}{\underset{\sim}{T}}_{S}$ in place of $\underset{\sim}{T}_{S}$, the equations of motion, in the present situation, become quite involved; we remark only that they may be reduced to the global form

$$\overset{\kappa}{\Sigma}{}^{iAB}_{k}\left(\frac{\partial^2 x^k}{\partial X^A \partial X^B} - \overset{\kappa}{\Gamma}{}^{C}_{AB}\frac{\partial x^k}{\partial X^C}\right) \tag{III-42}$$

$$- \overset{\kappa}{\Sigma}{}^{iA}\,\overset{\kappa}{T}{}^{B}_{AB} + \overset{\kappa}{\Sigma}{}^{i} + \rho_{\kappa} b^{i} = \rho_{\kappa}\ddot{x}^{i}$$

where the fields $\overset{\kappa}{\Sigma}{}^{iAB}_{k}$, $\overset{\kappa}{\Sigma}{}^{iA}$, and $\overset{\kappa}{\Sigma}{}^{i}$ are defined on $\mathcal{B}$; the $\overset{\kappa}{T}{}^{A}_{BC}$ are the components of the same torsion tensor which appeared before in equation (II-33). For the formal definitions of the (global) fields appearing in (III-42) the reader may consult Wang ([32], §4); his notation differs only slightly from that which we have employed here.

The field equations of motion (III-41) have been derived under the restrictive assumption that our generalized elastic body $\mathcal{B}$ is in a single phase only, say, $P(\mathcal{B})$. If the set $M(\mathcal{B})$ of §4 consists of more than one phase then a <u>phase transition</u> may take place during the course of a motion $\phi(t): \mathcal{B} \to \mathcal{R}^3$. Some interesting examples of the kinds of phase transition which can be dealt with within the theory of smooth materially uniform generalized elastic bodies are discussed in §5 of Wang [32]; in addition, the general theory of this chapter enables us to consider the problem of a change of homogeneity in inhomogeneous simple elastic bodies. For example, suppose that $\mathcal{B}$ is a simple thermoelastic body which is undergoing some thermodynamical process and that θ_i and θ_f are the initial and final index sections for $\mathcal{B}$ in such a process (i.e., we are looking at example (ii) §2). Suppose that when we hold the index section fixed at either θ_i or θ_f, the body behaves just like the smooth materially uniform elastic bodies of Chapter II, i.e., if $p,q \in \mathcal{B}$ then exist isomorphisms

$$\underset{\sim}{r}_i(p,q): \mathcal{B}_p \to \mathcal{B}_q$$

$$\underset{\sim}{r}_f(p,q): \mathcal{B}_p \to \mathcal{B}_q$$

such that for all $\kappa: \mathcal{B}_p \to R^3$,

$$\underset{\sim}{E}_p(\underset{\sim}{\kappa},\theta_i(p)) \equiv \underset{\sim}{E}_q(\underset{\sim}{\kappa} \circ \underset{\sim}{r}_i(p,q)^{-1}, \theta_i(q)) \qquad \text{(III-43)}$$

and

$$\underset{\sim}{E}_p(\underset{\sim}{\kappa}, \theta_f(p)) \equiv \underset{\sim}{E}_q(\underset{\sim}{\kappa} \circ \underset{\sim}{r}_f(p,q)^{-1}, \theta_f(q)) \qquad \text{(III-44)}$$

If in the neighborhood of any $p \in \mathcal{B}$ we can find smooth local fields $\underset{\sim}{r}_i(p,\cdot)$ and $\underset{\sim}{r}_f(p,\cdot)$ satisfying (III-43) and (III-44), respectively, then, provided θ_i and θ_f may be held fixed, we can suppress the dependence of the response functions on the index sections and regard $\mathcal{B}$, in each case, as a smooth materially uniform elastic body of the kind dealt with in Chapter II; the inhomogeneity of $\mathcal{B}$, in each of these cases, will differ, in general, as in one case it depends on the field $\underset{\sim}{r}_i(p,\cdot)$ and in other on the field $\underset{\sim}{r}_f(p,\cdot)$, for points $p\varepsilon\mathcal{B}$. The fields $\underset{\sim}{r}_i(p,\cdot)$ and $\underset{\sim}{r}_f(p,\cdot)$ need not be related, of course, in any definite way. An alternative approach to the problem of a change of inhomogeneity, within the context of the anelasticity theory of Eckart [15] is presented in the next chapter.

Chapter IV. Anelastic Behavior and Dislocation Motion

1. Introduction

Like the theories of material uniformity presented in Chapters II and III, the geometric structure theory for anelastic materials, which will be presented here, is based on the concept of material isomorphism. For anelastic bodies, however, the inhomogeneity, in each given motion, is not permanently fixed in the body manifold (as we shall soon see, this is a consequence of the fact that the material atlas which characterizes the distribution of the response functions on the body manifold is now time-dependent). Our aim, therefore, in this chapter is to describe, as completely as possible, the evolution of inhomogeneity in anelastic materials of the type first considered by Eckart [15]. Stated in more direct terms, we will equip the points of our body manifold $\mathcal{B}$ with constitutive relations similar to those first proposed by Eckart and then study the evolution of the geometric structures which arise from the concept of material isomorphism that is naturally associated with the consitutive relations of the points comprising $\mathcal{B}$.

Prior to stating the constitutive relations for points of an anelastic body $\mathcal{B}$, we need to append to the material on kinematics (Chapter II, §2) thus far presented, the following(8)

(8) local configurations of $p\varepsilon\mathcal{B}$ will be denoted here by $\underset{\sim}{r}(p)$ instead of $\underset{\sim}{r}_p$, as in chapter II.

Definition III-1 Let $\phi(t): \mathcal{B} \to R^3$ be a motion of $\mathcal{B}$, $\underset{\sim}{r}(p,t): \mathcal{B}_p \to R^3$ a local motion of p, and $t_o > 0$ a fixed time. Then the one parameter family $\phi_s^{t_o} \equiv \phi(t_o - s)$, $s\varepsilon[0,\infty)$, is called the history of $\mathcal{B}$ up to the time t_o in the motion ϕ. In a similar manner, we call $\underset{\sim}{r}_s^{t_o}(p) \equiv \underset{\sim}{r}(p,t_o - s)$ the local history of $p\varepsilon\mathcal{B}$, up to time t_o, in the local motion $\underset{\sim}{r}(p,t)$.

For $s>0$, $\phi(t_o - s)$ is termed the past configuration at time-lapse s from the present time. If $\phi(t): \mathcal{B} \to R^3$ is a motion of $\mathcal{B}$ then we say that the history of $\mathcal{B}$ up to the two distinct times t_o, t_1 is the same if $\phi_s^{t_o} = \phi_s^{t_1}$, $\forall s\varepsilon[0,\infty)$. If the motion ϕ has constant history ϕ_o then $\phi(t) = \phi_o$, $\forall t\varepsilon R$ and a history $\phi_s^{t_o}$ such that $\phi_s^{t_o} = \phi_o$, $\forall s\varepsilon[0,\infty)$, is called a rest history (at ϕ_o). Naturally, $\underset{\sim}{r}_s^{t_o}(p)$ is called a local rest history (at $\underset{\sim}{r}_o(p): \mathcal{B}_p \to R^3$) if $\underset{\sim}{r}_s^{t_o}(p) = \underset{\sim}{r}_o(p)$, $\forall s\varepsilon[0,\infty)$. If $\underset{\sim}{F}: R^3 \to R^3$ is a deformation gradient we shall denote the local deformation history via $\underset{\sim}{F}^{t_o}(s) \equiv \underset{\sim}{F}(t_o - s)$, $\forall s\varepsilon[0,\infty)$.

2. Elastic and Anelastic Response Functions; Anelastic Transformations

The class of anelastic materials, which was first considered in a systematic way by Eckart in 1948 [15], form a very special subclass of a larger group of materials which can be termed quasi-elastic. The concept of a quasi-elastic material point was introduced by Wang & Bowen [33] in 1966 in an attempt to formulate a thermodynamical theory for non-linear materials with memory which was more general than that which was first proposed by Coleman [34], [35] for simple materials with fading memory; in this work of Wang & Bowen the need to trace the local configurations of particles $p\varepsilon\mathcal{B}$ back to past infinity (i.e., as $s\to\infty$), and to specify the memory effect of the history of each state variable, is obviated. The complete thermodynamical structure presented in [33] shall, however, not be needed in our work here as we will be dealing, once again, with a purely mechanical theory. (Material uniformity, actually symmetry uniformity, within the context of thermoelasticity, and the thermodynamics of dislocation motions, will be treated in the next chapter). We can, therefore, begin with the following

Definition III-2. Let $\mathcal{B}$ be a material body. A particle $p\varepsilon\mathcal{B}$ is called a quasi-elastic point if at any time t_o in any local motion $\underset{\sim}{r}(p,t)$ of p we can define an instantaneous response function $\underset{\sim}{E}_p^{t_o}$ such that the stress tensor $\underset{\sim}{T}_S$ in the present local configuration $\underset{\sim}{r}(p,t_o)$ is given by

$$\underset{\sim}{T}_S = \underset{\sim}{E}_p^{t_o}(\underset{\sim}{r}(p,t_o)) = \underset{\sim}{E}_p^{t_o}(\underset{\sim}{r}_o^{t_o}(p)) \qquad \text{(IV-1)}$$

Remark The response function $\underset{\sim}{E}_p^{t_o}$ at time t_o is determined, in general, by the past history $\underset{\sim}{r}_s^{t_o}(p)$, $s\varepsilon(0,\infty)$, of p in the local motion up to time t_o. Thus, if the past histories $\underset{\sim}{r}_s^{t_o}(p)$ and $\underset{\sim}{r}_s^{t_1}(p)$ of p up to the distinct times t_o, t_1, in the local motion $\underset{\sim}{r}(p,t)$, are the same (i.e., if they agree on $(0,\infty)$, we will have $\underset{\sim}{E}_p^{t_o}(\underset{\sim}{r}(p)) = \underset{\sim}{E}_p^{t_1}(\underset{\sim}{r}(p))$ for all local configurations $\underset{\sim}{r}(p)$: $\mathcal{B}_p \to \mathcal{R}^3$ of p; we note also that if $\underset{\sim}{r}(p,t_o) = \underset{\sim}{r}(p,t_1)$ then the stress tensor at p at these two distinct times will also be the same. On the other hand, if $\hat{r}(p,t)$ and $r(p,t)$, $t\varepsilon\mathcal{R}$, are two different local motions of $p\varepsilon\mathcal{B}$, then, in general, $\hat{\underset{\sim}{r}}_s^{t_o}(p) \neq \underset{\sim}{r}_s^{t_o}(p)$ for an arbitrary $t_o\varepsilon\mathcal{R}$; the instanteous response functions corresponding to these two local motions will then also be different. We can, therefore, conclude that while an instantaneous response function $\underset{\sim}{E}_p^{t_o}$ can be defined at each $t_o\varepsilon\mathcal{R}$, in each local motion of $p\varepsilon\mathcal{B}$, $E_p^{t_o}$ is <u>not</u> a <u>fixed</u> <u>material</u> <u>function</u> of p at t_o.

In order to single out from the class of quasi-elastic materials (bodies comprised soley of quasi-elastic material points) that important subclass of materials which may be identified as being <u>anelastic</u>, in the sense of Eckart [15], we follow Wang & Bloom [19] and lay down the following assumptions: each $p\varepsilon\mathcal{B}$ is a quasi-elastic point and

I. For each $p\varepsilon\mathcal{B}$, there exists on <u>elastic</u> <u>response</u> <u>function</u> $\underset{\sim}{E}_p$ defined on $\mathcal{D}_p$ (the set of all local configurations $\underset{\sim}{r}(p)$ of p) such that the stress tensor $\underset{\sim}{T}_s^o$ in any rest history at $\underset{\sim}{r}_o\varepsilon\mathcal{D}_p$ is given by

$$\underset{\sim}{T}_s^o = \underset{\sim}{E}_p(\underset{\sim}{r}_o) \qquad \text{(IV-2)}$$

For each $p\varepsilon\mathcal{B}$, $\underset{\sim}{E}_p$ is a smooth function on $\mathcal{D}_p$ whose values lie in $L_s(R^3,R^3)$.

II. In any local motion $\underset{\sim}{r}(p,t)$ of p (i.e., p is not at rest) there exists an <u>anelastic</u> <u>transformation</u> <u>function</u> $\underset{\sim}{\alpha}(p,t): B_p \to B_p$, $\forall t\varepsilon R$, with the properties

(i) $\underset{\sim}{\alpha}(p,t)$ is a smooth function of t whose values are linear orientation-preserving automorphisms of B_p for each $t\varepsilon R$

(ii) $\underset{\sim}{\alpha}(p,t) \to id_{B_p}$ as $t\to-\infty$

(iii) the <u>instantaneous</u> (<u>anelastic</u>) <u>response</u> <u>function</u> $\underset{\sim}{E}_p^{t_o}$ of p at time t_o is given via

$$\underset{\sim}{E}_p^{t_o}(\underset{\sim}{r}(p)) = \underset{\sim}{E}_p(\underset{\sim}{r}(p)\circ\underset{\sim}{\alpha}(p,t_o)) \qquad \text{(IV-3)}$$

where $\underset{\sim}{r}(p)$ is any element of $\mathcal{D}_p$.

<u>Remark</u> As $\underset{\sim}{E}_p$ is a smooth function on $\mathcal{D}_p$ and $\underset{\sim}{\alpha}(p,t)$ is a smooth function of t, clearly, $\underset{\sim}{E}_p^t(\underset{\sim}{r}(p))$ depends smoothly on both t and $\underset{\sim}{r}(p)$. Also, by virtue of condition (ii) above, we have

$$\underset{\sim}{E}_p^t(\underset{\sim}{r}(p)) \to \underset{\sim}{E}_p(\underset{\sim}{r}(p)), \ t\to-\infty \qquad \text{(IV-4)}$$

Note also that, as a consequence of (IV-3), we have $\underset{\sim}{E}_p^t(\underset{\sim}{r}(p)\circ\underset{\sim}{\alpha}(p,t)^{-1}) = \underset{\sim}{E}_p(\underset{\sim}{r}(p))$, $\forall t\varepsilon R$; thus the stress tensor in the local configuration $\underset{\sim}{r}(p)\circ\underset{\sim}{\alpha}(p,t)^{-1}$ is independent of t, in each fixed local configuration $\underset{\sim}{r}(p)$ of p, for all $t\varepsilon R$. As it is usually assumed in continuum mechanics that the stress tensors in two local configurations of the same material point cannot be the same, unless the densities in those local configurations are the same, we will also impose assumption

(iv) in any local motion $\underset{\sim}{r}(p,t)$ of $p \varepsilon \mathcal{B}$ an anelastic transformation function $\underset{\sim}{\alpha}(p,t)$ must be isochoric, i.e., $\underset{\sim}{\alpha}(p,t) \varepsilon SL(\mathcal{B}_p)$, $\forall t \varepsilon \mathcal{R}$.

We now choose a fixed local reference configuration for p which we will designate by $\underset{\sim}{r}(p)$, i.e., $\underset{\sim}{r}(p)$ will no longer denote any arbitrary element of $\mathcal{D}_p$ but, rather, a fixed element which we have singled out from $\mathcal{D}_p$. As in Chapters II and III, if $\phi\colon \mathcal{B} \to \mathcal{R}^3$ is any configuration of $\mathcal{B}$, then we can represent the local configuration ϕ_{*p} in $\mathcal{D}_p$ by the deformation gradient $\underset{\sim}{F}(\equiv \phi_{*p} \circ \underset{\sim}{r}(p)^{-1})$ from $r(p)$ to ϕ_{*p}. Obviously, we may define a <u>relative elastic response function</u> $\underset{\sim}{S}_{\underset{\sim}{r}(p)}$ in a manner similar to that of Chapter II, i.e., $\forall \underset{\sim}{F} \varepsilon GL(3)$

$$\underset{\sim}{S}_{\underset{\sim}{r}(p)}(\underset{\sim}{F},p) \equiv \underset{\sim}{E}_p(\underset{\sim}{F} \circ \underset{\sim}{r}(p)) \tag{IV-5}$$

and, if $\underset{\sim}{r}(p,t)$ is any local motion of p, we can also define a <u>relative anelastic response function</u> via

$$\underset{\sim}{S}^t_{\underset{\sim}{r}(p)}(\underset{\sim}{F},p) \equiv \underset{\sim}{E}^t_p\,(\underset{\sim}{F} \circ \underset{\sim}{r}(p)),\ \forall t \varepsilon \mathcal{R},\ \underset{\sim}{F} \varepsilon GL(3) \tag{IV-6}$$

Now, by putting (IV-3), (IV-5), and (IV-6) together we obtain, for any $t_o \varepsilon \mathcal{R}$,

$$\underset{\sim}{S}^{t_o}_{\underset{\sim}{r}(p)}(\underset{\sim}{F}) = \underset{\sim}{S}_{\underset{\sim}{r}(p)}(\underset{\sim}{F}\underset{\sim}{A}_{\underset{\sim}{r}(p)}(t_o)),\ \forall \underset{\sim}{F} \varepsilon GL(3) \tag{IV-7}$$

as the relationship between $S^{t_o}_{\underset{\sim}{r}(p)}$ and $S_{\underset{\sim}{r}(p)}$, where

$$\underset{\sim}{A}_{\underset{\sim}{r}(p)}(t) \equiv \underset{\sim}{r}(p) \circ \underset{\sim}{\alpha}(p,t) \circ \underset{\sim}{r}(p)^{-1}, \quad \forall t \varepsilon R$$

is called the <u>relative anelastic transformation function</u>. Clearly, $\underset{\sim}{A}_{\underset{\sim}{r}(p)}(t)$ is a smooth tensor-valued function which is isochoric, for each $t\varepsilon R$, and satisfies $\underset{\sim}{A}_{\underset{\sim}{r}(p)}(t) \to \underset{\sim}{1}$, as $t \to -\infty$; as a consequence, we see that $S^t_{\underset{\sim}{r}(p)}(\underset{\sim}{F})$ depends smoothly on both $\underset{\sim}{F}$ and t and, in addition, $S^t_{\underset{\sim}{r}(p)}(\underset{\sim}{F}) \to S_{\underset{\sim}{r}(p)}(\underset{\sim}{F})$ as $t \to -\infty$. The analysis above also goes through, of course, for elements of $\mathcal{D}_p$ which are not the induced local configuration of any global configuration ϕ of $\mathcal{B}$.

3. <u>Anelastic Symmetry Groups and Anelastic Inner Products</u>.

The various response functions and relative response functions introduced in the previous section give rise, of course, to a variety of symmetry groups and to a set of relationships among these groups and the anelastic transformations functions $\underset{\sim}{\alpha}(p,t)$ and $\underset{\sim}{A}_{\underset{\sim}{r}(p)}(t)$. To begin with, we define the <u>elastic symmetry group</u> $g(p)$ via

$$g(p) = \{\underset{\sim}{\xi}\varepsilon SL(\mathcal{B}_p) \,|\, \underset{\sim}{E}_p(\underset{\sim}{\delta} \circ \underset{\sim}{\xi}) = \underset{\sim}{E}_p(\underset{\sim}{\delta}), \quad \forall \underset{\sim}{\delta} \varepsilon \mathcal{D}_p\}$$

and the <u>anelastic symmetry group</u> $g^t(p)$ by

$$g^t(p) = \{\underset{\sim}{\xi}\varepsilon SL(\mathcal{B}_p) \,|\, \underset{\sim}{E}^t_p(\underset{\sim}{\delta} \circ \underset{\sim}{\xi}) = \underset{\sim}{E}^t_p(\underset{\sim}{\delta}), \quad \forall \underset{\sim}{\delta} \varepsilon \mathcal{D}_p\}$$

An immediate consequence of these definitions and (IV-3)

is the relation

$$g^t(p) = \underset{\sim}{\alpha}(p,t) \circ g(p) \circ \underset{\sim}{\alpha}(p,t)^{-1} \qquad \text{(IV-8)}$$

Note also that if $\underset{\sim}{\alpha}(p,t)$ is an anelastic transformation function in the local motion $\underset{\sim}{r}(p,t)$ of $p\varepsilon B$ then so is $\bar{\underset{\sim}{\alpha}}(p,t)$ provided that $\underset{\sim}{\alpha}(p,t)^{-1} \circ \bar{\underset{\sim}{\alpha}}(p,t) \equiv \underset{\sim}{\xi}(t) \varepsilon g(p)$, $\forall t\varepsilon R$, and $\underset{\sim}{\xi}(t) \to id_{B_p}$ as $t\to-\infty$. Conversely, if $\underset{\sim}{\xi}(t)$ is a smooth function with $\underset{\sim}{\xi}(t) \varepsilon g(p)$, $\forall t\varepsilon R$ and $\underset{\sim}{\xi}(t) \to id_{B_p}$, as $t\to-\infty$ then $\bar{\alpha}(p,t) \equiv \alpha(p,t) \circ \underset{\sim}{\xi}(t)$ is an anelastic transformation function in the local motion $\underset{\sim}{r}(p,t)$ if $\underset{\sim}{\alpha}(p,t)$ is. Thus, if $g(p)$ is non-trivial, (IV-3) and the requirement that $\underset{\sim}{\alpha}(p,t) \to id_{B_p}$ as $t\to-\infty$ do not uniquely determine an anelastic transformation function in a given local motion of p. From (IV-8) and the fact that $\underset{\sim}{\alpha}(p,t) \to id_{B_p}$ as $t\to-\infty$ we deduce that $g^t(p)$ depends smoothly on t and, moreover, $g^t(p) \to g(p)$ as $t\to-\infty$. The reader can easily check that (IV-8) is independent of the choice of $\underset{\sim}{\alpha}(p,t)$ within the equivalence class of anelastic transformations defined by $\alpha(p,t)^{-1} \circ \bar{\alpha}(p,t) \equiv \underset{\sim}{\xi}(t) \varepsilon g(p)$, $\forall t\varepsilon R$ and $\underset{\sim}{\xi}(t) \to id_{B_p}$ as $t\to-\infty$.

The <u>relative elastic symmetry group</u> $G_{\underset{\sim}{r}(p)}$ is defined, as usual, by

$$G_{\underset{\sim}{r}(p)} \equiv \{\underset{\sim}{K}\varepsilon SL(3) \mid S_{\underset{\sim}{r}(p)}(\underset{\sim}{F}\underset{\sim}{K},p) = S_{\underset{\sim}{r}(p)}(\underset{\sim}{F},p),\ \forall \underset{\sim}{F}\varepsilon GL(3)\}$$

and the <u>relative anelastic symmetry group</u> $G^t_{\underset{\sim}{r}(p)}$ is given by

$$G^t_{\underset{\sim}{r}(p)} \equiv \{\underset{\sim}{K}\varepsilon SL(3) | S^t_{\underset{\sim}{r}(p)}(\underset{\sim}{F}\underset{\sim}{K},p) = S^t_{\underset{\sim}{r}(p)}(\underset{\sim}{F},p),\ \forall \underset{\sim}{F}\varepsilon GL(3)\}$$

Concerning the relationship which exists between $\mathcal{G}_{\underset{\sim}{r}(p)}$ and $G^t_{\underset{\sim}{r}(p)}$ we note, first of all, that

$$\mathcal{G}_{\underset{\sim}{r}(p)} = \underset{\sim}{r}(p) \circ g(p) \circ \underset{\sim}{r}(p)^{-1}, \qquad \text{(IV-11)}$$

as in Chapter II, and that

$$G^t_{\underset{\sim}{r}(p)} = \underset{\sim}{r}(p) \circ g^t(p) \circ \underset{\sim}{r}(p)^{-1} \qquad \text{(IV-12)}$$

It then follows, at once, from (IV-8), that

$$G^t_{\underset{\sim}{r}(p)} = A_{\underset{\sim}{r}(p)}(t) \circ \mathcal{G}_{\underset{\sim}{r}(p)} \circ A_{\underset{\sim}{r}(p)}(t)^{-1} \qquad \text{(IV-13)}$$

so that $G^t_{\underset{\sim}{r}(p)} \to \mathcal{G}_{\underset{\sim}{r}(p)}$ as $t\to-\infty$; the relations (IV-11) and (IV-12) are simple consequences of the definitions of the various symmetry groups involved and the relationships which exist among their associated response functions. Note also that just as $\underset{\sim}{\alpha}(p,t)$ is non-uniquely determined, in general, if $g(p)$ is non-trivial so will $\underset{\sim}{A}_{\underset{\sim}{r}(p)}(t)$ be non-uniquely determined, i.e., if $\underset{\sim}{A}_{\underset{\sim}{r}(p)}(t)$ is a relative anelastic transformation function so is $\bar{A}_{\underset{\sim}{r}(p)}(t) \equiv A_{\underset{\sim}{r}(p)}(t) \circ \underset{\sim}{K}(t)$ provided $\underset{\sim}{K}(t) \varepsilon \mathcal{G}_{\underset{\sim}{r}(p)}$, $\forall t\varepsilon R$, and $\underset{\sim}{K}(t) \to \underset{\sim}{1}$ as $t\to-\infty$.

The definitions of the symmetry groups given above

enable us to apply Noll's general classification of materials to anelastic material points. For instance, we will call p an <u>anelastic fluid point</u> if $\mathcal{G}(p) = SL(\mathcal{B}_p)$; in view of (IV-11) this is equivalent to stating that $G_{\underset{\sim}{r}(p)} = SL(3)$ relative to any local reference configuration $\underset{\sim}{r}(p)$ of p. The following theorem concerning fluid points has an almost trivial proof:

<u>Theorem</u> IV-1 Every anelastic fluid point is an elastic fluid point.

<u>Proof</u>: To begin with let us note that $p \in \mathcal{B}$ is an elastic fluid point if the symmetry group $\mathcal{G}^t(p)$ of the response function $\underset{\sim}{E}_p^t$ coincides with $SL(\mathcal{B}_p)$ for each $t \in R$. So, let r(p,t) be a local motion of p for which $\underset{\sim}{\alpha}(p,t)$ is an associated anelastic transformation function. Then $\bar{\underset{\sim}{\alpha}}(p,t) = \underset{\sim}{\alpha}(p,t) \circ \xi(t)$ will also be an anelastic transformation function in the local motion $\underset{\sim}{r}(p,t)$ if $\underset{\sim}{\xi}(t) \in g \equiv SL(\mathcal{B}_p)$ for each $t \in R$ and $\underset{\sim}{\xi}(t) \to id_{\mathcal{B}p}$ as $t \to -\infty$; as $\underset{\sim}{\alpha}(p,t)$ is isochoric, i.e., $\underset{\sim}{\alpha}(p,t) \in SL(\mathcal{B}_p)$, $\forall t \in R$, and $\underset{\sim}{\alpha}(p,t) \to id_{\mathcal{B}p}$ as $t \to -\infty$, we may choose $\underset{\sim}{\xi}(t) \equiv \underset{\sim}{\alpha}(p,t)^{-1}$. Thus, $\bar{\alpha}(p,t) \equiv id_{\mathcal{B}_p}$, $\forall t \in R$, is an anelastic transformation function in the local motion r(p,t) of p and, therefore, by (IV-3)

$$\underset{\sim}{E}_p^t(\underset{\sim}{\delta}(p)) = E_p(\underset{\sim}{\delta}(p)), \quad \forall t \in R, \quad \forall \underset{\sim}{\delta} \in \mathcal{D}_p$$

$$\Rightarrow \mathcal{G}^t(p) = \mathcal{G}(p) \equiv SL(\mathcal{B}_p), \quad \forall t \in R$$

which completes the proof. <u>Q. E. D.</u>

In order to define the concept of an <u>anelastic solid point</u> we must assume that $\mathcal{B}_p$ can be equipped with an intrinsic elastic inner product m such that $g(p) \subset \mathcal{O}(\mathcal{B}_p)^{(9)}$. In other words if $\underset{\sim}{\xi} \varepsilon g(p)$ imples that

$$m(\underset{\sim}{\xi}\underset{\sim}{u}, \underset{\sim}{\xi}\underset{\sim}{v}) = m(\underset{\sim}{u}, \underset{\sim}{v}), \quad \underset{\sim}{u}, \underset{\sim}{v} \varepsilon \mathcal{B}_p. \tag{IV-15}$$

then $g(p) \subset \mathcal{O}(\mathcal{B}_p)$ and p is called an <u>anelastic solid point</u>. If $g(p) \equiv \mathcal{O}(\mathcal{B}_p)$, relative to some inner product m on $\mathcal{B}_p$, then p is called an <u>isotropic anelastic solid point</u>. As in Chapter II, we will call a local reference configuration $\underset{\sim}{r}(p)$ <u>undistorted</u> if $G_{\underset{\sim}{r}(p)} \subset \mathcal{O}(3)$; if such is the case then $\underset{\sim}{K} \varepsilon G_{\underset{\sim}{r}(p)} \Rightarrow \underset{\sim}{K}\underset{\sim}{U} \cdot \underset{\sim}{K}\underset{\sim}{V} = \underset{\sim}{U} \cdot \underset{\sim}{V}$, $\forall \underset{\sim}{U}, \underset{\sim}{V} \varepsilon \mathcal{R}^3$. In view of the relationship (IV-8) between the symmetry groups $g(p)$ and $g^t(p)$ we can define an intrinsic <u>anelastic inner product</u> m^t on $\mathcal{B}_p$ via

$$m^t(\underset{\sim}{u}, \underset{\sim}{v}) \equiv m(\underset{\sim}{\alpha}^{-1}(p,t)\underset{\sim}{u}, \underset{\sim}{\alpha}^{-1}(p,t)\underset{\sim}{v}), \quad \forall \underset{\sim}{u}, \underset{\sim}{v} \varepsilon \mathcal{B}_p \tag{IV-16}$$

and each $t \varepsilon \mathcal{R}$ in each local motion $\underset{\sim}{r}(p,t)$ for which $\underset{\sim}{\alpha}(p,t)$ is an associated anelastic transforamtion function. It follows that if $g(p) \subset \mathcal{O}(\mathcal{B}_p)$ relative to m, $g^t(p) \subset \mathcal{O}(\mathcal{B}_p)$ relative to m^t for each $t \varepsilon \mathcal{R}$, i.e.,
$\underset{\sim}{\xi} \varepsilon g^t(p) \Rightarrow m^t(\underset{\sim}{\xi}\underset{\sim}{u}, \underset{\sim}{\xi}\underset{\sim}{v}) = m^t(\underset{\sim}{u}, \underset{\sim}{v})$, $\forall \underset{\sim}{u}, \underset{\sim}{v} \varepsilon \mathcal{B}_p$.

<u>Exercise</u> 16 If m^t is defined by (IV-16) verify that $g(p) \subset \mathcal{O}(\mathcal{B}_p)$ relative to m $\Rightarrow g^t(p) \subset \mathcal{O}(\mathcal{B}_p)$ relative to m^t (for each $t \varepsilon \mathcal{R}$). Show also that if $\underset{\sim}{r}(p)$ is undistorted,

(9) $\mathcal{O}(\mathcal{B}_p)$ is the rotation group on $\mathcal{B}_p$ relative to m

$G^t_{\underset{\sim}{r}(p)} \subset O(3)$ relative to the inner product $(\ ,\ )^t$ on R^3 which is defined by

$$(\underset{\sim}{U},\underset{\sim}{V})^t \equiv \underset{\sim}{A}^{-1}_{\underset{\sim}{r}(p)}(t)\underset{\sim}{U} \cdot \underset{\sim}{A}^{-1}_{\underset{\sim}{r}(p)}(t)\underset{\sim}{V}, \quad \forall \underset{\sim}{U},\underset{\sim}{V} \in R^3 \qquad \text{(IV-17)}$$

(Note that both m^t as defined by (IV-16) and $(\ ,\ )^t$ as defined by (IV-17) depend smoothly on t with $m^t(\underset{\sim}{u},\underset{\sim}{v}) \to m(\underset{\sim}{u},\underset{\sim}{v})$ and $(U,V)^t \to \underset{\sim}{U}\cdot\underset{\sim}{V}$, as $t \to -\infty$.

4. Flow Rules; Uniqueness of the Anelastic Transformation Function

Suppose that $\underset{\sim}{\alpha}(p,t)$ is an anelastic transformation function in some local motion $\underset{\sim}{r}(p,t)$; as we have already indicated in §3, $\underset{\sim}{\alpha}(p,t)$, which is determined by $\underset{\sim}{r}^t_s(p)$, is generally non-unique if $g(p)$ is non-trivial. In any application of the theory of this chapter, we would assume that $\underset{\sim}{\alpha}(p,t)$ is governed by a flow rule of the form $\dot{\underset{\sim}{\alpha}} = \underset{\sim}{\psi}$ where $\underset{\sim}{\psi}$ is some functional of the past local histories $\underset{\sim}{r}^t_s(p)$ of the anelastic point p. While we will not employ any special form for $\underset{\sim}{\psi}$ in this treatise we will impose the condition that $\underset{\sim}{\psi}$ be a smooth function of t in each local motion $\underset{\sim}{r}(p,t)$. Of course, we must append to the flow rule the "initial" condition $\underset{\sim}{\alpha}(p,t) \to id_{Bp}$ as $t \to -\infty$.

Because $\underset{\sim}{\alpha}$ is not unique, at each time t in a given local motion $\underset{\sim}{r}(p,t)$ a value of $\underset{\sim}{\psi}$ must be assigned at some $\underset{\sim}{\alpha}(p,t)$; our choices are limited, of course, to the equivalence class engendered by the relation $\underset{\sim}{\alpha}(p,t)^{-1} \circ \bar{\underset{\sim}{\alpha}}(p,t) \equiv \underset{\sim}{\xi}(t) \in g(p)$

$\forall t \in R$, with $\underset{\sim}{\xi}(t) \to id_{B_p}$ as $t \to -\infty$. To make matters a bit worse, we note that if $\underset{\sim}{\alpha}$ and $\underset{\sim}{\bar{\alpha}}$ belong to the same equivalence class and $\xi(t_o) = id_{B_p}$, for some time t_o, then generally $\dot{\underset{\sim}{\bar{\alpha}}}(p,t_o) \neq \dot{\underset{\sim}{\alpha}}(p,t_o)$, even though $\underset{\sim}{\alpha}(p,t_o) = \underset{\sim}{\bar{\alpha}}(p,t_o)$. In fact

$$\dot{\underset{\sim}{\bar{\alpha}}}(p,t_o) - \dot{\underset{\sim}{\alpha}}(p,t_o) = \underset{\sim}{\alpha}(p,t_o)\underset{\sim}{\nu} \qquad \text{(IV-18)}$$

where $\underset{\sim}{\nu} \in \underset{\sim}{g}_p$, the Lie algebra of $g(p)$. Thus, $\underset{\sim}{\psi}(\underset{\sim}{\alpha}(p,t_o), t_o)$ is unique only to within an arbitrary additive tensor in $\underset{\sim}{\alpha}(p,t_o)\underset{\sim}{g}_p$.

The non-uniqueness of $\underset{\sim}{\psi}(\underset{\sim}{\alpha}(p,t_o), t_o)$ described above may be removed in the following way: note first of all that because $\underset{\sim}{\alpha}(p,t) \in SL(B_p)$, $\forall t \in R$, we must have $\underset{\sim}{\alpha}(p,t_o)^{-1} \circ \dot{\underset{\sim}{\alpha}}(p,t_o) \in SL(B_p)$ or

$$tr(\underset{\sim}{\alpha}(p,t_o)^{-1} \circ \underset{\sim}{\psi}(\underset{\sim}{\alpha}(p,t_o), t_o)) = 0 \qquad \text{(IV-19)}$$

for each $t_o \in R$; but this implies that $\underset{\sim}{\psi}(\underset{\sim}{\alpha}(p,t_o), t_o)$ must belong to $\underset{\sim}{\alpha}(p,t_o) \circ sl(B_p)$. If we decompose $sl(B_p)$ into the direct sum

$$sl(B_p) = \underset{\sim}{g}_p \oplus \underset{\sim}{h}_p \qquad \text{(IV-20)}$$

then we may define a unique value for $\underset{\sim}{\psi}(\underset{\sim}{\alpha}(p,t_o), t_o)$ by imposing the <u>selection condition</u>:

$$\underset{\sim}{\alpha}(p,t_o)^{-1} \circ \underset{\sim}{\psi}(\underset{\sim}{\alpha}(p,t_o), t_o) \in \underset{\sim}{h}_p \qquad \text{(IV-21)}$$

Now that we have found a condition which enables us to uniquely determine the value of $\dot{\alpha}(p,t_o)$, we turn our attention to a theorem which guarantees us that a unique solution of the flow rule exists which satisfies the inital condition and the selection condition (IV-21).

Theorem IV-2 Let h_p be the subspace of $SL(\mathcal{B}_p)$ which is given by (IV-20) and define $[ad(\xi)]\eta \equiv \xi \circ \eta \circ \xi^{-1}$ for elements $\xi, \eta \varepsilon L(\mathcal{B}_p, \mathcal{B}_p)$. Then if h_p is invariant under the adjoint representation $ad(\xi)$ for all $\xi \varepsilon \hat{g}(p)$, the identity component of $g(p)$, the solution of the flow rule $\dot{\alpha} = \psi$, subject to the initial condition, ($\alpha(p,t) \to id_{\mathcal{B}p}$ as $t \to -\infty$) and the selection condition (IV-21), is uniquely determined.

Proof Let $\alpha(p,t)$ and $\bar{\alpha}(p,t)$ be any two solutions of $\dot{\alpha} = \psi$ which lie in the same equivalence class, i.e., $\alpha(p,t)^{-1} \circ \bar{\alpha}(p,t) \equiv \xi(t)$ where $\xi(t) \varepsilon\, g(p)$, $\forall t \varepsilon R$ and $\xi(t) \to id_{\mathcal{B}p}$ as $t \to -\infty$. Differentiating with respect to t and using the fact that(10) $\frac{d}{dt}(\alpha^{-1}) = -\alpha^{-1} \circ \dot{\alpha} \circ \alpha^{-1}$ we obtain

$$\dot{\xi} = -\alpha^{-1} \circ \dot{\alpha} \circ \alpha^{-1} \circ \bar{\alpha} + \alpha^{-1} \circ \dot{\bar{\alpha}} \qquad \text{(IV-22)}$$

or

$$-\xi^{-1} \circ \alpha^{-1} \circ \psi(\alpha,\cdot) \circ \xi + \bar{\alpha}^{-1} \circ \psi(\bar{\alpha},\cdot) = \xi^{-1} \circ \dot{\xi} \qquad \text{(IV-23)}$$

(10) to avoid long expressions we now drop the displayed dependence of α on p and t and assume this to be understood.

as $\dot{\underset{\sim}{\alpha}}(p,\cdot) = \underset{\sim}{\psi}(\underset{\sim}{\alpha}(p,\cdot), \cdot)$ and $\dot{\bar{\underset{\sim}{\alpha}}}(p,\cdot) = \underset{\sim}{\psi}(\bar{\underset{\sim}{\alpha}}(p,\cdot), \cdot)$. By using the definition of "adjoint representation" which is given in the statement of the theorem, we may rewrite (IV-23) in the form

$$-[ad(\underset{\sim}{\xi})](\underset{\sim}{\alpha}^{-1} \circ \underset{\sim}{\psi}(\underset{\sim}{\alpha},\cdot)) + \underset{\sim}{\alpha}^{-1} \circ \underset{\sim}{\psi}(\underset{\sim}{\alpha},\cdot) = \underset{\sim}{\xi}^{-1} \circ \dot{\underset{\sim}{\xi}} \qquad \text{(IV-24)}$$

By virtue of the hypothesis of the theorem and the selection condition (IV-21), the left hand side of (IV-24) is contained in $\underset{\sim}{h}_p$; but $\underset{\sim}{\xi}^{-1} \circ \dot{\underset{\sim}{\xi}} \in g_p$ since $\underset{\sim}{\xi}(t) \in \mathcal{g}(p)$, $\forall t \in R$. Since $\underset{\sim}{h}_p$ is, by definition, the orthogonal complement of $\underset{\sim}{g}_p$ in $sl(\mathcal{B}_p)$, it must follow that $\underset{\sim}{\xi}^{-1} \circ \dot{\underset{\sim}{\xi}} = \underset{\sim}{0}$ so $\underset{\sim}{\xi}(t) = \underset{\sim}{\xi}_o$, the initial value of $\underset{\sim}{\xi}$, for all $t \in R$. As $\underset{\sim}{\xi}(t) \to id_{\mathcal{B}p}$ for $t \to -\infty$, we must have $\underset{\sim}{\xi}_o = id_{\mathcal{B}p}$; this, in turn, implies that $\underset{\sim}{\xi}(t) \equiv id_{\mathcal{B}p}$, $\forall t \in R$, or $\underset{\sim}{\alpha}(p,t) = \bar{\underset{\sim}{\alpha}}(p,t)$, $\forall t \in R$. Q. E. D.

For solid and fluid anelastic points it is a relatively simple matter to verify that the orthogonal complements in $sl(\mathcal{B}_p)$ of the Lie algebras of the respective isotropy groups $\mathcal{g}(p)$ are invariant under the action of $ad(\underset{\sim}{\xi})$ for all $\underset{\sim}{\xi} \in \hat{\mathcal{g}}(p)$. We know, in fact, that for an anelastic fluid point p, $\mathcal{g}(p) = sl(\mathcal{B}_p)$ so $\underset{\sim}{g}_p = sl(\mathcal{B}_p)$; this implies that $\underset{\sim}{h}_p = \underset{\sim}{0}$ and that the unique solution of $\dot{\underset{\sim}{\alpha}}(t,p) = \underset{\sim}{0}$ satisfying $\underset{\sim}{\alpha}(t,p) \to id_{\mathcal{B}_p}$ as $t \to -\infty$ must be, of course, $\underset{\sim}{\alpha}(t,p) \equiv id_{\mathcal{B}_p}$, $\forall t \in R$. On the other hand, if p is an anelastic solid point then we may choose an intrinsic elastic m on $\mathcal{B}_p$ such that $\mathcal{g}(p) \subset \mathcal{O}(\mathcal{B}_p)$ relative to m. Now let $\underset{\sim}{\alpha}$

and $\underset{\sim}{\beta}$ be two transformations in $L(\mathcal{B}_p;\mathcal{B}_p)$ and let $\underset{\sim}{\beta}^T$ denote the transpose of $\underset{\sim}{\beta}$ with respect to m. We define the inner product on $L(\mathcal{B}_p;\mathcal{B}_p)$, which is induced by m, via

$$(\underset{\sim}{\alpha},\underset{\sim}{\beta}) \equiv tr(\underset{\sim}{\alpha}\circ\underset{\sim}{\beta}^T), \quad \forall\underset{\sim}{\alpha},\underset{\sim}{\beta}\varepsilon L(\mathcal{B}_p,\mathcal{B}_p)$$

Now let $\underset{\sim}{h}_p$ be the orthogonal complement of $\underset{\sim}{g}_p$ in $sl(\mathcal{B}_p)$, relative to the inner product (,) on $L(\mathcal{B}_p,\mathcal{B}_p)$, i.e., $(\underset{\sim}{\alpha},\underset{\sim}{\beta}) = \underset{\sim}{0}$, $\forall\underset{\sim}{\alpha}\varepsilon\underset{\sim}{h}_p$, $\underset{\sim}{\beta}\varepsilon\underset{\sim}{g}_p$. To show that $\underset{\sim}{h}_p$ is invariant under $ad(\underset{\sim}{\xi})$, for all $\underset{\sim}{\xi}\varepsilon\hat{g}(p)$, we choose $\underset{\sim}{\xi}\varepsilon L(\mathcal{B}_p,\mathcal{B}_p)$ such that $\underset{\sim}{\xi}$ is an orthogonal transformation relative to m; we then have $\underset{\sim}{\xi}^{-1} = \underset{\sim}{\xi}^T$ and for $\underset{\sim}{\alpha}\varepsilon\underset{\sim}{h}_p$, $\underset{\sim}{\beta}\varepsilon\underset{\sim}{g}_p$

$$\begin{aligned}([ad(\underset{\sim}{\xi})]\underset{\sim}{\alpha},\ \underset{\sim}{\beta}) &= tr(\underset{\sim}{\xi}\circ\underset{\sim}{\alpha}\circ\underset{\sim}{\xi}^{-1}\circ\underset{\sim}{\beta}^T)\\ &= tr(\underset{\sim}{\alpha}\circ(\underset{\sim}{\xi}^{-1}\circ\underset{\sim}{\beta}\circ\underset{\sim}{\xi})^T)\\ &= (\underset{\sim}{\alpha},\ [ad(\underset{\sim}{\xi}^{-1})]\underset{\sim}{\beta})\\ &= 0\end{aligned}$$

since $\underset{\sim}{g}_p$ is invariant under $ad(\underset{\sim}{\xi}^{-1})$ for all $\underset{\sim}{\xi}\varepsilon\hat{g}(p)$. It now follows at once, from the definition of $\underset{\sim}{h}_p$ as the orthogonal complement of $\underset{\sim}{g}_p$ in $sl(\mathcal{B}_p)$, that $[ad(\underset{\sim}{\xi})]\underset{\sim}{\alpha}\ \varepsilon\ \underset{\sim}{h}_p$ if $\alpha\varepsilon\underset{\sim}{h}_p$. We deduce, from the theorem above, that for an anelastic solid point p the anelastic transformation function $\underset{\sim}{\alpha}(p,t)$, $\forall t\varepsilon\mathcal{R}$, is unique in each local motion $\underset{\sim}{r}(p,t)$ of p if $\dot{\underset{\sim}{\alpha}}$ satisfies the selection condition, namely, if $[\underset{\sim}{\alpha}(p,t)^{-1} \circ \dot{\underset{\sim}{\alpha}}(p,t)] \perp \underset{\sim}{g}_p$, $\forall t\varepsilon\mathcal{R}$, relative to any intrinsic elastic inner product on $\mathcal{B}_p$.

Now suppose that we define the relative anelastic transformation $A_{\underset{\sim}{r}(p)}(t)$ by $A_{\underset{\sim}{r}(p)}(t) \equiv \underset{\sim}{r}(p) \circ \underset{\sim}{\alpha}(p,t) \circ \underset{\sim}{r}(p)^{-1}$, $\forall t \varepsilon R$, where $\underset{\sim}{r}(p)$ is our local reference configuration. Then the flow rule $\dot{\underset{\sim}{\alpha}}(p,t) = \underset{\sim}{\psi}(\alpha(p,t), t)$ has the representation $\dot{\underset{\sim}{A}}_{\underset{\sim}{r}(p)}(t) = \underset{\sim}{\Psi}_{\underset{\sim}{r}(p)}(\underset{\sim}{A}_{\underset{\sim}{r}(p)}(t), t)$ and the initial condition involved is $\bar{A}_{\underset{\sim}{r}(p)}(t) \to \underset{\sim}{1}$, as $t \to -\infty$. As in [19] we now suppress the subscript $\underset{\sim}{r}(p)$ in $\underset{\sim}{A}_{\underset{\sim}{r}(p)}$, the local reference configuration $\underset{\sim}{r}(p)$ being understood; then within the equivalence class defined by $\underset{\sim}{A}(t)^{-1}\bar{\underset{\sim}{A}}(t) \equiv \underset{\sim}{K}(t) \varepsilon G_{\underset{\sim}{r}(p)}$, $\forall t \varepsilon R$, and the initial condition $\underset{\sim}{K}(t) \to \underset{\sim}{1}$ as $t \to -\infty$, the value of $\underset{\sim}{\Psi}(\underset{\sim}{A}(t), t)$ is unique to within an arbitrary additive tensor belonging to $A(t)\underset{\sim}{G}_p$ where $\underset{\sim}{G}_p$ denotes the Lie algebra of $G_{\underset{\sim}{r}(p)}$. If $\underset{\sim}{H}_p$ denotes the orthogonal complement of $\underset{\sim}{G}_p$ in $sl(3)$, i.e., $sl(3) = \underset{\sim}{G}_p \oplus \underset{\sim}{H}_p$ then the selection rule, in this case, assumes the form

$$\underset{\sim}{A}(t)^{-1}\underset{\sim}{\Psi}(\underset{\sim}{A}(t), t) \;\varepsilon\; \underset{\sim}{H}_p, \quad \forall t \varepsilon R \qquad \text{(IV-25)}$$

<u>Remark</u> If p is an anelastic solid point and $\underset{\sim}{r}(p)$ is an undistorted local reference configuration of p then the selection rule (IV-25) becomes

$$tr(\underset{\sim}{A}(t)^{-1}\underset{\sim}{\Psi}(\underset{\sim}{A}(t), t)\underset{\sim}{K}) = 0, \quad \forall t \varepsilon R, \quad \forall \underset{\sim}{K} \varepsilon \underset{\sim}{G}_p$$

If this condition is satisfied then A(t) is unique in each

local motion of p.

Now consider $\underset{\sim}{S}(\underset{\sim}{F},p)$ and $\underset{\sim}{S}^t(F,p)$, $\underset{\sim}{F}\varepsilon GL(3)$, the elastic and anelastic response functions, respectively, relative to the fixed local reference configuration $\underset{\sim}{r}(p)$ (the $\underset{\sim}{r}(p)$ subscripts on $\underset{\sim}{S}$, $\underset{\sim}{S}^t$ have been dropped). By (IV-7) the relationship between $\underset{\sim}{S}$ and $\underset{\sim}{S}^t$ is given by $\underset{\sim}{S}^t(\underset{\sim}{F},p) = \underset{\sim}{S}(\underset{\sim}{F}\underset{\sim}{A}(t), p)$ which can be written in component form relative to the standard basis of $\mathcal{R}^3$ as

$$S^a_b(F^c_d, t) = S^a_b(F^c_f\ A^f_d(t)) \qquad \text{(IV-26)}$$

where t has been written as an explicit argument and the dependence of the response functions on the point $p\varepsilon\mathcal{B}$ has not been displayed. If we set $F^c_f = \delta^c_f$ in (IV-26), differentate with respect to t, and use the flow rule, we get

$$\dot{T}^a_b(t) = S^{ad}_{bc}(A^e_f(t))\Psi^c_d(A^e_f(t), t) \qquad \text{(IV--27)}$$

where $T^a_b(t) \equiv S^a_b(A^c_d(t))$ are the components of the stress tensor in the given local reference configuration and $S^{ad}_{bc}(F^e_f) \equiv \partial S^a_b(F^e_f)/\partial F^c_d$, $\forall \underset{\sim}{F}\varepsilon GL(3)$. At any fixed value of $A^e_f(t)$, therefore, $\dot{T}^a_b(t)$ depends linearly on the flow rate $\Psi^c_d(A^e_f(t), t)$. While (IV-27) represents a necessary condition to be satisfied by $\underset{\sim}{\Psi}$, this condition alone in is not, in general, sufficient to determine $\underset{\sim}{\Psi}$.

Exercise 17 Show that when p is an anelastic solid crystal

point, i.e., $G = \{\underset{\sim}{1}\}$, $\underset{\sim}{G}_p = \{\underset{\sim}{0}\}$ and $\underset{\sim}{H}_p = sl(3)$, relative to an undistorted local configuration $\underset{\sim}{r}(p)$, (IV-27) can not be solved uniquely for Ψ^c_d in terms of $\dot{T}^a_b$.

Before concluding this section, we should point out that there are exceptions to the type of situation illustrated by the example in the above exercise, i.e., if p is an isotropic anelastic solid point the system of equations equivalent to (IV-27) may be invertible. In order to see this we let $\underset{\sim}{r}(p)$ be an undistorted reference configuration of p and define an elastic inner product m on $\mathcal{B}_p$ via

$$m(\underset{\sim}{u},\underset{\sim}{v}) = \underset{\sim}{r}(p)\underset{\sim}{u} \cdot \underset{\sim}{r}(p)\underset{\sim}{v}, \quad \forall \underset{\sim}{u},\underset{\sim}{v} \in \mathcal{B}_p \tag{IV-27}$$

Using (IV-16) we define the anelastic inner product m^t on $\mathcal{B}_p$ by

$$\begin{aligned} m^t(\underset{\sim}{u},\underset{\sim}{v}) &= m(\underset{\sim}{\alpha}(p,t)^{-1}\underset{\sim}{u},\underset{\sim}{\alpha}(p,t)^{-1}\underset{\sim}{v}) \qquad \text{(IV-28)} \\ &= \underset{\sim}{r}(p)\circ\underset{\sim}{\alpha}(p,t)^{-1}\underset{\sim}{u} \cdot \underset{\sim}{r}(p)\circ\underset{\sim}{\alpha}(p,t)^{-1}\underset{\sim}{v} \\ &= \underset{\sim}{A}_{\underset{\sim}{r}(p)}(t)^{-1}\underset{\sim}{r}(p)\underset{\sim}{u} \cdot \underset{\sim}{A}_{\underset{\sim}{r}(p)}(t)^{-1}\underset{\sim}{r}(p)\underset{\sim}{v} \\ &= \underset{\sim}{r}(p)\underset{\sim}{u} \cdot \underset{\sim}{C}^{-1}(t)\underset{\sim}{r}(p)\underset{\sim}{v} \end{aligned}$$

where $\underset{\sim}{C}(t) \equiv A_{\underset{\sim}{r}(p)}(t)A^T_{\underset{\sim}{r}(p)}(t)$ is called the left Cauchy-Green tensor of $A_{\underset{\sim}{r}(p)}$. If we define $\underset{\sim}{\gamma}(t) \equiv \underset{\sim}{\alpha}(p,t)\circ\underset{\sim}{\alpha}(p,t)^T$ to be the left Cauchy-Green tensor of $\underset{\sim}{\alpha}(p,t)$, relative to m, then it is a simple matter to show that

$$m^t(\underset{\sim}{u},\underset{\sim}{v}) = m(\underset{\sim}{u}, \underset{\sim}{\gamma}(t)^{-1}\underset{\sim}{v}) \tag{IV-29}$$

$$\underset{\sim}{C}(t) = \underset{\sim}{r}(p) \circ \underset{\sim}{\gamma}(t) \circ \underset{\sim}{r}(p)^{-1},$$

i.e., $\underset{\sim}{C}(t)$ is the representation of $\underset{\sim}{\gamma}(t)$, relative to the local reference configuration $\underset{\sim}{r}(p)$. Moreover, $\underset{\sim}{\gamma}(t)$ and $\underset{\sim}{C}(t)$ are smooth functions of t in each local motion $\underset{\sim}{r}(p,t)$ and, clearly, $\underset{\sim}{\gamma}(t) \to id_{\mathcal{B}_p}$ and $\underset{\sim}{C}(t) \to \underset{\sim}{1}$, as $t\to-\infty$. It is a relatively simple matter to show that $\underset{\sim}{\alpha}(p,t)$ and $\underset{\sim}{\bar{\alpha}}(p,t)$ are anelastic transformations which belong to the same equivalence class (in the sense, of course, that $\underset{\sim}{\alpha}(p,t)^{-1} \circ \bar{\alpha}(p,t)\varepsilon g(p)$, $\forall t\varepsilon R$ and $\underset{\sim}{\alpha}(p,t)^{-1} \circ \bar{\alpha}(p,t) \to id_{\mathcal{B}_p}$ as $t\to-\infty$) iff, $\forall t\varepsilon R$, $\underset{\sim}{\alpha}(p,t) \circ \underset{\sim}{\alpha}(p,t)^T = \bar{\alpha}(p,t) \circ \bar{\alpha}(p,t)^T$, i.e., iff $\underset{\sim}{\gamma}(t) = \underset{\sim}{\bar{\gamma}}(t)$.

<u>Exercise</u> 18 Verify the above statement concerning anelastic transformations belonging to the same equivalence class. Show also that $A_{\underset{\sim}{r}(p)}(t)$ and $\bar{A}_{\underset{\sim}{r}(p)}(t)$ satisy $A_{\underset{\sim}{r}(p)}(t) \circ A_{r(p)}(t)^T = \bar{A}_{\underset{\sim}{r}(p)}(t) \circ \bar{A}_{\underset{\sim}{r}(p)}(t)^T$, $\forall t\varepsilon R$, iff these relative anelastic transformation functions satisfy $A^{-1}_{\underset{\sim}{r}(p)}(t) \circ \bar{A}_{\underset{\sim}{r}(p)}(t) \equiv \underset{\sim}{K}(t) \;\varepsilon\; G_{\underset{\sim}{r}(p)}$, $\forall t\varepsilon R$, and $\underset{\sim}{K}(t) \to \underset{\sim}{1}$ as $t\to-\infty$.

As the exercise above indicates, the respective equivalence classes of $\underset{\sim}{\alpha}(p,t)$ and $A_{\underset{\sim}{r}(p)}(t)$ are completely determined by $\underset{\sim}{\gamma}(t)$ and $\underset{\sim}{C}(t)$. Thus, we may replace the flow rules

$$\dot{\underset{\sim}{\alpha}} = \underset{\sim}{\psi} \quad \underline{\text{and}} \quad \dot{\underset{\sim}{A}} = \underset{\sim}{\Psi}$$

by flow rules of the form

$$\dot{\underset{\sim}{\gamma}} = \underset{\sim}{\pi} \quad \text{and} \quad \dot{\underset{\sim}{C}} = \underset{\sim}{\Pi}$$

where $\underset{\sim}{\pi} = \pi^T$, $\underset{\sim}{\Pi} = \Pi^T$ and (to insure the uniqueness of $\underset{\sim}{\gamma}(t)$ and $\underset{\sim}{C}(t)$ at each $t\varepsilon R$) $tr(\underset{\sim}{\gamma}^{-1} \circ \underset{\sim}{\pi}) = 0$, $tr(\underset{\sim}{C}^{-1} \circ \underset{\sim}{\Pi}) = 0$.

<u>Exercise</u> 19 Show that $\underset{\sim}{\pi}$ and $\underset{\sim}{\Pi}$ determine the values of $\underset{\sim}{\psi}$ and $\underset{\sim}{\Psi}$ via

$$\underset{\sim}{\psi}(\underset{\sim}{\alpha}(p,t),\ t) = \tfrac{1}{2}\underset{\sim}{\pi}(t) \circ [\underset{\sim}{\alpha}(p,t)^{-1}]^T \qquad \text{(IV-30)}$$

$$\underset{\sim}{\Psi}(\underset{\sim}{A}(t),\ t) = \tfrac{1}{2}\underset{\sim}{\Pi}(t) \circ [\underset{\sim}{A}(t)^{-1}]^T$$

Since it can be shown [5] that the relative response function $\underset{\sim}{S}(\underset{\sim}{F},p)$ of an isotropic solid can be represented by a function $\hat{\underset{\sim}{S}}(\underset{\sim}{D},p)$, of the left-Cauchy Green tensor $\underset{\sim}{D}$ of $\underset{\sim}{F}$, if the local reference configuration $\underset{\sim}{r}(p)$ is undistorted, we can now reduce the system of equations $T^a_b(t) = S^a_b(A^c_d(t))$ to the system $T^a_b(t) = \hat{S}^a_b(C^e_f(t))$; differentiation with respect to t now yields

$$\dot{T}^a_b(t) = \hat{S}^{ad}_{bc}(C^e_f(t))\Pi^c_d(t) \qquad \text{(IV-31)}$$

where $\hat{S}^{ag}_{bh}(C^e_f) \equiv \partial\hat{S}^a_b(C^e_f)/\partial C^h_g$ for all positive-definite symmetric linear transformations of R^3. Because the values of $\underset{\sim}{\Pi}$ belong to a linear space of dimension 5 while those of $\underset{\sim}{\dot{T}}$ belong to a space of dimension (not greater than) 6 it may now be possible to invert (IV-31) and solve for Π^c_d in terms of $\dot{T}^a_b$; this contrasts sharply with the situation which exists relative to the system (IV-27) where, in general, $\underset{\sim}{\dot{T}}$ belongs to a space of dimension 6 whereas $\underset{\sim}{\Psi}$ takes its values in sl(3) which is of dimension 8.

5. Material Uniformity in the Theory of Anelasticity

Let B be an anelastic body, i.e., a material body consisting soley of anelastic points; as the key to the development of a geometric theory for such materials rests with the concepts of elastic and anelastic materials isomorphisms, we need to proceed with some definitions. So, let $p,q \varepsilon B$. Then we can state

Definition IV-3 A linear orientation-preserving isomorphism $\underset{\sim}{r}(p,q)$: $B_p \to B_q$ is called an elastic material isomorphism of p and q if

$$\underset{\sim}{E}_p(\underset{\sim}{\nu}(q) \circ \underset{\sim}{r}(p,q)) = \underset{\sim}{E}_q(\underset{\sim}{\nu}(q)), \quad \forall \underset{\sim}{\nu}(q) \varepsilon D_q \tag{IV-32}$$

If an isomorphism $\underset{\sim}{r}(p,q)$ satisfying (IV-32) exists we say that p and q are materially isomorphic and, furthermore, if the points of B are pairwise materially isomorphic then we say that our anelastic body B is materially uniform;

this definition is completely consistent with the definition of material uniformity presented in Chapter II, as the elastic response function $\underset{\sim}{E}_p$ is the response function of $p\varepsilon\mathcal{B}$ in any local rest history at some $\underset{\sim}{\nu}(p)\varepsilon\mathcal{D}_p$. It is very easy to see now that relative to the elastic response functions $\underset{\sim}{E}_p$, $p\varepsilon\mathcal{B}$, many of our results in Chapter II may be carried over without change. For instance, the isotropy groups $\mathcal{G}(p)$ and $\mathcal{G}(q)$ associated, respectively, with $\underset{\sim}{E}_p$ and $\underset{\sim}{E}_q$ are related via (II-8) where $\underset{\sim}{r}(p,q)$ now denotes any <u>elastic material isomorphism</u> of p and q. Now let us suppose that the anelastic body $\mathcal{B}$ is not at rest but, rather, that $\mathcal{B}$ is undergoing some motion $\phi(t)$ which induces a local motion $\phi_{*p}(t)$ of each $p\varepsilon\mathcal{B}$; then the response of $p\varepsilon\mathcal{B}$, at any time $t\varepsilon R$, in the local motion $\phi_{*p}(t)$, is characterized by the anelastic response function $\underset{\sim}{E}_p^t$, which is related to $\underset{\sim}{E}_p$ by (IV-3) with $\underset{\sim}{\alpha}(p,t)$ an anelastic transformation function (for the local motion $\phi_{*p}(t)$). We can, therefore, make the following

<u>Definition</u> IV-4 Fix $t\varepsilon R$. An isomorphism $\underset{\sim}{r}^t(p,q)\colon \mathcal{B}_p \rightarrow \mathcal{B}_q$ is called an <u>anelastic material isomorphism</u> of p and q, <u>at time t</u>, in the motion $\phi(t)$ of $\mathcal{B}$, if

$$\underset{\sim}{E}_p^t(\underset{\sim}{\nu}(q)\circ\underset{\sim}{r}^t(p,q)) = \underset{\sim}{E}_q^t(\underset{\sim}{\nu}(q)), \quad \forall\underset{\sim}{\nu}(q)\varepsilon\mathcal{D}_q \tag{IV-33}$$

If we compare (IV-33) with (IV-32) and make use of the relationship (IV-3) between the response functions $\underset{\sim}{E}_p$ and

$\underset{\sim}{E}_p^t$, at time t, we can easily prove

Theorem IV-3 In any motion $\phi(t)$ of $\mathcal{B}$, $\underset{\sim}{r}^t(p,q)$ is an anelastic material isomorphism of p and q, at time t, iff

$$\underset{\sim}{r}(p,q) \equiv \underset{\sim}{\alpha}(q,t)^{-1} \circ \underset{\sim}{r}^t(p,q) \circ \underset{\sim}{\alpha}(p,t) \qquad \text{(IV-34)}$$

is an elastic material isomorphism of p and q (note that $\underset{\sim}{\alpha}(\cdot,t)$ is an anelastic transformation function in the local motion $\phi_{*(\cdot)}(t)$).

It is a direct consequence of theorem IV-3 that an anelastic body $\mathcal{B}$ is materially uniform if and only if it is anelastically materially uniform at any one time (and hence at all times $t \in \mathcal{R}$) in any one motion (and hence in all motions $\phi(t)\colon \mathcal{B} \to \mathcal{R}^3$); by saying that $\mathcal{B}$ is anelastically materially uniform (at time t in the motion $\phi(t)$ of $\mathcal{B}$) we mean that for any pair of points $p,q \in \mathcal{B}$, there exists an anelastic material isomorphism $\underset{\sim}{r}^t(p,q)$ satisfying (IV-33).

Exercise 20 By making use of II-8, with $\underset{\sim}{r}(p,q)$ an elastic material isomorphism, show that the relationship (IV-34) is independent of the choice of $\underset{\sim}{\alpha}(p,t)$ and $\underset{\sim}{\alpha}(q,t)$ within their respective equivalence classes.

Exercise 21 If $\underset{\sim}{r}^t(p,q)$ is any anelastic material isomorphism of p and q, at time t, in the motion $\phi(t)$ of $\mathcal{B}$, show that

$$\mathcal{g}^t(q) = \underset{\sim}{r}^t(p,q) \circ \mathcal{g}^t(p) \circ \underset{\sim}{r}^t(p,q)^{-1} \qquad \text{(IV-35)}$$

We will now add the following basic assumption about $\mathcal{B}$ to those already given [11]:

III. the anelastic body $\mathcal{B}$ is materially uniform and for each $p \in \mathcal{B}$ there exists a neighborhood $\mathcal{N}_p$ and a <u>smooth field</u> $\underset{\sim}{r}(p,\cdot)$ of elastic material isomorphisms defined on $\mathcal{N}_p$; we do not require that $\mathcal{N}_p \equiv \mathcal{B}$ for any $p \in \mathcal{B}$

<u>Exercise</u> 22 Show that if $\underset{\sim}{r}(p,q)$ is a given elastic material isomorphism of p and q and $\underset{\sim}{\bar{r}}(p,q)$ is any linear isomorphism for $\mathcal{B}_p$ to $\mathcal{B}_q$ then $\underset{\sim}{\bar{r}}(p,q)$ is also an elastic material isomorphism of p and q iff $\underset{\sim}{\bar{r}}(p,q)^{-1} \circ \underset{\sim}{r}(p,q) \in g(p)$ (or iff $\underset{\sim}{r}(p,q)^{-1} \circ \underset{\sim}{\bar{r}}(p,q) \in g(q)$). Thus, an elastic material isomorphism of p and q is unique iff $g(p) \equiv \{id_{\mathcal{B}_p}\}$; note also that by virtue of (II-8), with $\underset{\sim}{r}(p,q)$ an elastic material isomorphism of p and q, $g(q)$ is the trivial group $id_{\mathcal{B}_q}$ if $g(p)$ is the trivial group $id_{\mathcal{B}_q}$.

<u>Remarks</u> The results of the above exercise point to the fact that it is generally impossible to extend a smooth local field of elastic material isomorphism, say $\underset{\sim}{r}(p,\cdot)$ defined on $\mathcal{N}_p$, to a smooth global field on $\mathcal{B}$. Exceptions do exist, i.e., every smooth local field of elastic material isomorphisms on a star-shaped body $\mathcal{B}$ can be extended to a smooth global field of material isomorphisms; for further details we refer the reader to the discussion in Wang [7].

(11) the "smoothness" part of this assumption is consistent with the one set down in Chapter II for (smooth) materially uniform elastic bodies.

Now let us note that because of our third assumption (above) concerning smooth materially uniform anelastic bodies the stress tensor field $\underset{\sim}{T}_S$ must be smooth in any smooth rest history of $\mathcal{B}$; if we want the same thing to be true at each time t in any smooth motion $\phi(t)$ of $\mathcal{B}$ then it is clear, in view of (IV-3), that we must add assumption

IV. In each smooth motion $\phi(t)$ of $\mathcal{B}$ an anelastic transformation function $\underset{\sim}{\alpha}(p,\cdot)$ can be chosen, for each $p\varepsilon\mathcal{B}$, in such a way that $\underset{\sim}{\alpha}(\cdot,t)$ is a smooth (global) function on $\mathcal{B}$ for each $t\varepsilon\mathcal{R}$.

One very important thing to note, concerning this last assumption, is that it requires of $\underset{\sim}{\alpha}(\cdot,t)$, for each $t\varepsilon\mathcal{R}$, that it be a smooth global field defined on $\mathcal{B}$ and not just a smooth local field on some neighborhood $\mathcal{N}_p$ of each $p\varepsilon\mathcal{B}$. Our reasons for imposing such a requirement here are quite simple: as $t\to-\infty$, $\alpha(p,t) \to id_{\mathcal{B}_p}$, for each $p\varepsilon\mathcal{B}$, and the identity map $id_{\mathcal{B}_{(\cdot)}}$ is certainly a smooth (global) field on $\mathcal{B}$. In fact, in view of our smoothness assumptions and equation (IV-34) we can consider the distribution of the elastic response functions $\underset{\sim}{E}_{(\cdot)}$ on $\mathcal{B}$ as inducing an <u>initial structure</u> on $\mathcal{B}$ from which the <u>anelastic structures</u> on $\mathcal{B}$, in any given motion $\phi(t)$, are generated smoothly in time via the associated field $\underset{\sim}{\alpha}(\cdot,t)$ of anelastic transformation functions; in other words the field of anelastic response functions $E^t_{(\cdot)}$ will be distributed smoothly on $\mathcal{B}$, at each time $t\varepsilon\mathcal{R}$, in any motion $\phi(t)$ of $\mathcal{B}$. To be more precise, let $\phi(t)$ be a motion of $\mathcal{B}$ and $\underset{\sim}{r}(p,\cdot)$ a smooth field of elastic material isomorphisms defined on some neighorhood N_p of p; then at each time t in the motion $\phi(t)$ the field

$$\underset{\sim}{r}^t(p,\cdot) \equiv \underset{\sim}{\alpha}(\cdot,t) \circ \underset{\sim}{r}(p,\cdot) \circ \underset{\sim}{\alpha}(p,t)^{-1}, \qquad \text{(IV-36)}$$

where $\underset{\sim}{\alpha}(\cdot,t)$ denotes the smooth global field of anelastic transformation functions, whose existence is guaranteed by assumption IV, is a smooth field of anelastic material isomorphisms defined on N_p.

We are now in a position to describe how elastic and anelastic reference atlases for $\mathcal{B}$ are constructed; as regards the former kind of atlas we already know that the theory of Chapter II can be carried over to the present situation. In other words, an elastic reference chart is a pair $(\mathcal{U},\underset{\sim}{r})$ consisting of an open set $\mathcal{U} \subset \mathcal{B}$ and a smooth field $\underset{\sim}{r}(\cdot)$ of local reference configurations defined on $\mathcal{U}$ such that relative to $\underset{\sim}{r}(p)$ the elastic response function $\underset{\sim}{S}_{\underset{\sim}{r}(p)}(\underset{\sim}{F},\underset{\sim}{p})$ is independent of p, $\forall p \in \mathcal{U}$.(12) We set

$$S_{\underset{\sim}{r}}(F) \equiv \underset{\sim}{S}_{\underset{\sim}{r}(p)}(\underset{\sim}{F},p), \quad \forall \underset{\sim}{F} \in GL(3),\ p \in \mathcal{U} \qquad \text{(IV-37)}$$

and call $\underset{\sim}{S}_{\underset{\sim}{r}}$ the elastic response function relative to

(12) in Bloom & Wang [19] such charts were called material charts; we rename them reference charts here so as to be consistent with the definitions presented in Chapter II and in the original paper of Wang [7].

$(\mathcal{U},\underset{\sim}{r})$; the relative elastic symmetry group of $(\mathcal{U},\underset{\sim}{r})$ is $\underset{\sim}{\mathcal{G}_r} \equiv \mathcal{G}_{\underset{\sim}{r}(p)}$, of course, $\forall p\varepsilon\mathcal{U}$. As in Chapter II, we call two elastic reference charts $(\mathcal{U},\underset{\sim}{r})$ and $(\bar{\mathcal{U}},\underset{\sim}{\bar{r}})$ <u>compatible</u> if $\underset{\sim}{S}_{\underset{\sim}{r}}(\underset{\sim}{F}) = \underset{\sim}{S}_{\bar{\underset{\sim}{r}}}(\underset{\sim}{F})$, $\forall \underset{\sim}{F}\varepsilon GL(3)$ and a maximal collection $\underset{\sim}{\Phi} = \{(\mathcal{U}_\alpha,\underset{\sim}{r}^\alpha),\ \alpha\varepsilon I\}$ of pairwise compatible elastic reference charts such that $\mathcal{B} = \bigcup_I \mathcal{U}_\alpha$ is called an <u>elastic reference atlas</u> on $\mathcal{B}$. If $\underset{\sim}{\Phi}$ is an elastic reference atlas on $\mathcal{B}$ then we set $\underset{\sim}{S}_{\underset{\sim}{\Phi}}(\underset{\sim}{F}) \equiv \underset{\sim}{S}_{\underset{\sim}{r}^\alpha}(\underset{\sim}{F})$, $\forall\alpha\varepsilon I$ and $\mathcal{G}_{\underset{\sim}{\Phi}} \equiv \mathcal{G}_{\underset{\sim}{r}^\alpha}$, $\forall\alpha\varepsilon I$; naturally, $\underset{\sim}{S}_{\underset{\sim}{\Phi}}$ is called the <u>elastic response function relative</u> to $\underset{\sim}{\Phi}$ and $\mathcal{G}_{\underset{\sim}{\Phi}}$ is termed the <u>relative elastic symmetry group</u> of $\underset{\sim}{\Phi}$.

<u>Remarks</u> From the theory presented in Chapter II it follows that every smooth materially uniform anelastic body $\mathcal{B}$ can be equipped with an elastic reference atlas $\underset{\sim}{\Phi}$. The relationship between any two elastic reference atlases $\underset{\sim}{\Phi}$ and $\hat{\underset{\sim}{\Phi}}$ on $\mathcal{B}$, as well as the various relationships which hold among the corresponding relative elastic response functions and relative elastic symmetry groups are delineated in Chapter II; in particular, we recall that $\underset{\sim}{K}\varepsilon\mathcal{G}_{\underset{\sim}{\Phi}}$ iff $\underset{\sim}{K}\mathcal{G}_{\underset{\sim}{\Phi}} = \mathcal{G}_{\underset{\sim}{\Phi}}$ in the set-theoretic sense.

It should be obvious, at this point, that the definitions of homogeneity and local homogeneity presented in Chapter II also carry over to the present situation, i.e., we say that $\mathcal{B}$ is <u>locally elastically homogeneous</u> if for each $p\varepsilon\mathcal{B}$ there exists a neighborhood $\mathcal{N}_p$ and a configuration

$\underset{\sim}{\kappa}$: $N_p \to R^3$ such that (N_p, κ_*) is an elastic reference chart. Similarly, B is elasticity homogeneous if there exists a configuration κ: $B \to R^3$ such that (B, κ_*) is a (global) elastic reference chart.

Having reviewed the essential ingredients which go into the make-up of an elastic reference atlas we are now in a position to describe the construction of an anelastic reference atlas for the smooth anelastically materially uniform body B. Thus, let $\phi(t)$: $B \to R^3$ be a motion of B and suppose that the response of any $p \in B$ in the motion ϕ is characterized by the anelastic response function $\underset{\sim}{E}^t(\cdot)$ which is defined on D_p, $\forall t \in R$ and each $p \in B$. If we fix $t \in R$, then a pair (U,$\underset{\sim}{r}(\cdot,t)$) consisting of an open set $U \subset B$ and smooth field of local reference configurations defined on U (at time t) is called an anelastic reference chart if the relative anelastic response function at time t, i.e., $S^t_{\underset{\sim}{r}(p,t)}(\underset{\sim}{F},p)$, is independent of p, $\forall p \in U$. In such a case we set

$$\underset{\sim}{S}^t_{\underset{\sim}{r}}(\underset{\sim}{F}) \equiv \underset{\sim}{S}^t_{\underset{\sim}{r}(p,t)}(F,p), \quad \forall p \in U \qquad \text{(IV-38)}$$

and we call $\underset{\sim}{S}^t_{\underset{\sim}{r}}$ the relative anelastic response function of (U,$\underset{\sim}{r}(\cdot,t)$); the relative anelastic symmetry group of (U,$\underset{\sim}{r}(\cdot,t)$) is defined, of course, by the condition that $G^t_{\underset{\sim}{r}} \equiv G^t_{\underset{\sim}{r}(p,t)}$, $\forall p \in U$. Concerning the connection between elastic and anelastic reference charts on B, and their corresponding relative response functions, we can now state

<u>Theorem</u> IV-4 If $(\mathcal{U},\underset{\sim}{r})$ is an elastic reference chart with relative elastic response function $\underset{\sim}{S}_{\underset{\sim}{r}}$ then $(\mathcal{U},\underset{\sim}{r}(\cdot,t))$ is anelastic reference chart with relative anelastic response function $\underset{\sim}{S}^{t}_{\underset{\sim}{r}} \equiv S_{\underset{\sim}{r}}$ (on GL(3)) iff the smooth field $\underset{\sim}{\alpha}(\cdot,t)$ defined on $\mathcal{U}$ by

$$\underset{\sim}{\alpha}(p,t) \equiv \underset{\sim}{r}(p,t)^{-1} \circ \underset{\sim}{r}(p), \quad \forall p\varepsilon\mathcal{U} \tag{IV-39}$$

is a field of anelastic transformations at time t.

<u>Proof</u> We will prove the theorem going in just one direction; the proof in the other direction will be left as an exercise for the reader. So, suppose that $(\mathcal{U},\underset{\sim}{r})$ is an elastic reference chart with relative elastic response function $\underset{\sim}{S}_{\underset{\sim}{r}}$ and that $\underset{\sim}{\alpha}(p,t) \equiv \underset{\sim}{r}(p,t)^{-1} \circ \underset{\sim}{r}(p)$, $\forall p\varepsilon\mathcal{U}$, defines a field of anelastic transformations at time t. We want to show that $(\mathcal{U},\underset{\sim}{r}(\cdot,t))$ is anelastic reference chart whose relative anelastic response function $S^{t}_{\underset{\sim}{r}}$ coincides with $\underset{\sim}{S}_{r}$ on GL(3). So, employing (IV-3), with $\underset{\sim}{r}(p) \rightarrow \underset{\sim}{F}\circ\underset{\sim}{r}(p,t)$, $\underset{\sim}{F}$ an arbitrary element of GL(3), we get

$$\begin{aligned} E^{t}_{p}(\underset{\sim}{F}\circ\underset{\sim}{r}(p,t)) &= E_{p}(\underset{\sim}{F}\circ\underset{\sim}{r}(p,t)\circ\underset{\sim}{\alpha}(p,t) \\ &= E_{p}(\underset{\sim}{F}\circ\underset{\sim}{r}(p,t)\circ\underset{\sim}{r}(p,t)^{-1}\circ\underset{\sim}{r}(p)) \\ &= E_{p}(\underset{\sim}{F}\circ\underset{\sim}{r}(p)) \overset{\text{def}}{=} \underset{\sim}{S}_{\underset{\sim}{r}(p)}(\underset{\sim}{F},p) \\ &= \underset{\sim}{S}_{\underset{\sim}{r}}(\underset{\sim}{F}) \end{aligned} \tag{IV-40}$$

But $E^t_p(\underset{\sim}{F}\circ \underset{\sim}{r}(p,t)) \overset{\text{def}}{=} S^t_{\underset{\sim}{r}(p,t)}(\underset{\sim}{F},p)$ so

$$S^t_{\underset{\sim}{r}(p,t)}(\underset{\sim}{F},p) = \underset{\sim}{S}_{r}(\underset{\sim}{F}), \quad \forall \underset{\sim}{F}\varepsilon GL(3) \tag{IV-41}$$

from which the required result is immediate. Q. E. D.

The compatibility condition for two anelastic reference charts $(U,\underset{\sim}{r}(\cdot,t))$ and $(\bar{U},\bar{r}(\cdot,t))$ is defined in analogy with the similar condition for elastic reference charts, namely, we require that $S^t_{\underset{\sim}{r}}(\underset{\sim}{F}) = S^t_{\underset{\sim}{\bar{r}}}(F)$, $\forall \underset{\sim}{F}\varepsilon GL(3)$. An <u>anelastic reference atlas</u> $\underset{\sim}{\Phi}(t)$ <u>of</u> B <u>at time</u> t (in the motion $\phi(t)$ of B) is then defined to be a maximal collection $\underset{\sim}{\Phi}(t) = \{(U_\alpha,\underset{\sim}{r}^\alpha(\cdot,t)),\ \alpha\varepsilon I\}$ of pairwise compatible anelastic reference charts. Naturally, we set $S^t_{\underset{\sim}{\Phi}} = S^t_{\underset{\sim}{r}^\alpha}$, $\forall\alpha\varepsilon I$ and $G^t_{\underset{\sim}{\Phi}} = G^t_{\underset{\sim}{r}^\alpha}$, $\forall\alpha\varepsilon I$; these are called, respectively, the <u>relative anelastic response function of</u> $\underset{\sim}{\Phi}(t)$ and the <u>relative anelastic symmetry group</u> of $\underset{\sim}{\Phi}(t)$. The following set of exercises concern themselves with the relationship between different anelastic reference atlases on B and the transformation rules which apply when we change atlases; they are the direct counterparts of the results which apply when we make a change from one elastic reference atlas to another.

<u>Exercise</u> 23 Show that at each time $t\varepsilon R$ in each motion $\phi(t)$ of B there exists an anelastic reference atlas $\underset{\sim}{\Phi}(t)$. [<u>hint</u>: use Theorem IV-4 and the existence theorem for

elastic reference atlases $\underset{\sim}{\Phi}$]

Exercise 24 Show that any two anelastic reference atlases $\underset{\sim}{\Phi}(t)$ and $\hat{\underset{\sim}{\Phi}}(t)$ are related by

$$\hat{\underset{\sim}{\Phi}}(t) = \underset{\sim}{K}\underset{\sim}{\Phi}(t) = \{(U_\alpha, \underset{\sim}{K}\circ\underset{\sim}{r}^\alpha(\cdot,t)), \quad \alpha\varepsilon I\}$$

for some $\underset{\sim}{K}\varepsilon GL(3)$; deduce the transformation laws for $\underset{\sim}{S}^t_{\underset{\sim}{\Phi}} \to \underset{\sim}{S}^t_{\hat{\underset{\sim}{\Phi}}}$ and $G^t_{\underset{\sim}{\Phi}} \to G^t_{\hat{\underset{\sim}{\Phi}}}$ as $\underset{\sim}{\Phi} \to \hat{\underset{\sim}{\Phi}}$.

Now, it is a direct consequence of the transformation rules in the above exercise that there exists a one-to-one correspondence between anelastic reference atlases $\underset{\sim}{\Phi}(t)$ at time t and their associated relative response functions $\underset{\sim}{S}^t_{\underset{\sim}{\Phi}}$. Since we already know that such a one-to-one correspondence exists between elastic reference atlases $\underset{\sim}{\Phi}$ on B and their associated relative response functions $\underset{\sim}{S}_{\underset{\sim}{\Phi}}$, in each motion $\phi(t)$ of B we can establish a one-to-one correspondence between elastic and anelastic reference atlases $\underset{\sim}{\Phi}$ and $\underset{\sim}{\Phi}(t)$ via the requirement that $S_{\underset{\sim}{\Phi}}$ and $S^t_{\underset{\sim}{\Phi}}$ agree on GL(3); in fact, by virtue of Theorem IV-4, if $\underset{\sim}{\Phi} = \{(U_\alpha, \underset{\sim}{r}^\alpha), \quad \alpha\varepsilon I\}$ is an elastic reference atlas on B, a corresponding anelastic reference atlas would be given by

$$\underset{\sim}{\Phi}(t) = \{(U_\alpha, \underset{\sim}{r}^\alpha(\cdot) \circ \underset{\sim}{\alpha}(\cdot,t)^{-1}), \quad \alpha\varepsilon I\} \qquad \text{(IV-42)}$$

where $\underset{\sim}{\alpha}(\cdot,t)$ is any particular smooth global field of anelastic transformation functions on B; of course, this correspondence will not be invariant under a change of the field $\underset{\sim}{\alpha}(\cdot,t)$. Note also that because of the initial condition satisfied by all anelastic transformation functions, $\underset{\sim}{r}^{\alpha}(\cdot,t) \equiv \underset{\sim}{r}^{\alpha}(\cdot) \circ \underset{\sim}{\alpha}(\cdot,t)^{-1} \to \underset{\sim}{r}^{\alpha}(\cdot)$, as $t\to-\infty$, for each $\alpha\varepsilon I$; in this sense, we have $\underset{\sim}{\Phi}(t) \to \underset{\sim}{\Phi}$, as $t\to-\infty$. From this point on, if we use the same notations $\underset{\sim}{\Phi}$ and $\underset{\sim}{\Phi}(t)$ to denote elastic and anelastic reference atlases on $\mathcal{B}$ we do so with the assumption that the charts in these respective atlases are related via (IV-42).

6. Elastic and Anelastic Material Connections

Let $\mathcal{B}$ be a smooth materially uniform anelastic body. By an elastic material connection on $\mathcal{B}$ we mean an affine connection H on $\mathcal{B}$ such that the parallel transports of the tangent spaces relative to H, along any curve $\lambda \subset \mathcal{B}$, are elastic material isomorphisms. It should be clear that all the material of Chapter II dealing with material connections on an elastic body may be carried over here to elastic material connections on an anelastic body; in particular, we can state that every smooth materially uniform anelastic body $\mathcal{B}$ can be equipped with an elastic material connection. Also, all the results of §6 of Chapter II, concerning the relationship between the properties of homogeneity and local-homogeneity and the existence of certain kinds of material connections on an elastic body $\mathcal{B}$, may be carried over to anelastic bodies (e.g., a necessary and sufficient condition for an anelastic solid body $\mathcal{B}$ to be locally elastically-homogeneous is the existence of a flat elastic material connection). If $\underset{\sim}{\Phi}$ is an elastic reference atlas, $(U,\underset{\sim}{r})$ a reference chart in $\underset{\sim}{\Phi}$, and $\underset{\sim}{F} = \kappa_* \circ \underset{\sim}{r}^{-1}$ the deformation gradient from $\underset{\sim}{r}$ to κ_* (where $\kappa: B \to R^3$ induces the coordinate system (X^A) on B) then, by virtue of Theorem II-3, the functions $\Gamma^A_{BC}(X^D)$ are the connection symbols of an elastic material connection H on B iff the matrices

$$[F^{-1A}_{\ \ C}(\frac{\partial F^C_B}{\partial X^D} + \Gamma^C_{ED}F^E_B)], \quad D = 1,2,3$$

are contained in $\underset{\sim}{g}_{\Phi}$, the Lie algebra of $G_{\underset{\sim}{\Phi}}$.

We now want to turn our attention to the concept of an <u>anelastic material connection</u> on B; this is defined to be an affine connection H(t) on B such that the parallel transports defined by H(t), along all curves $\lambda \subset B$, are anelastic material isomorphisms. To show that there exists an anelastic material connection H(t) on B, at each time $t \in R$, in each motion $\phi(t)$ of B, we may demonstrate that each elastic material connection H on B gives rise to unique anelastic material connection H(t); the converse of this statement is also true. Thus, let $\lambda = \lambda(\tau)$ (13) be a curve in B and let $\rho(\tau)$ denote the parallel transport along λ, from $\lambda(0)$ to $\lambda(\tau)$, $\tau \le T$, which is induced by some elastic material connection H. Also, let $\phi(t)$ be any motion of B and $\underset{\sim}{\alpha}(\cdot,t)$ a smooth field of anelastic transformation functions defined on B in the motion $\phi(t)$. If for each $t \in R$

(13) $0 \le \tau \le T$ for some $T > 0$.

we define a map $\rho^t(\tau)\colon \mathcal{B}_{\lambda(0)} \to \mathcal{B}_{\lambda(\tau)}$ by

$$\rho^t(\tau) \equiv \underset{\sim}{\alpha}(\lambda(\tau),\ t) \circ \rho(\tau) \circ \underset{\sim}{\alpha}(\lambda(0),\ t)^{-1} \qquad \text{(IV-43)}$$

then, by virtue of (IV-36), $\rho^t(\tau)$ is an anelastic material isomorphism; conversely, if H(t) is an affine connection on $\mathcal{B}$ such that the parallel transports along λ, from $\lambda(0)$ to $\lambda(\tau)$, $\tau \leq T$, are given by anelastic material isomorphisms $\rho^t(\tau)$ then the maps $\rho(\tau)$ defined by (IV-43) are elastic material isomorphisms. Thus, (IV-43) establishes a one-to-one correspondence between elastic and anelastic material connections on $\mathcal{B}$.

In order to determine the connection symbols of H(t) in terms of those of an associated elastic material connection H (and the components of the particular field of anelastic transformation functions $\underset{\sim}{\alpha}(\cdot,t)$ which appears in (IV-43)), we choose a reference configuration $\kappa = (X^A)$ and note that the maps $\rho(\tau)$ and $\rho^t(\tau)$ are characterized by the equations of parallel transport

$$\frac{dv^A(\tau)}{d\tau} + \Gamma^A_{BC}(\lambda^D(\tau))v^B(\tau)\frac{d\lambda^C(\tau)}{d\tau} = 0 \qquad \text{(IV-44)}$$

and

$$\frac{du^A(\tau)}{d\tau} + \Gamma^A_{BC}(\lambda^D(\tau),\ t)u^B(\tau)\frac{d\lambda^C(\tau)}{d\tau} = 0 \qquad \text{(IV-45)}$$

where the components of $\lambda(\tau)$ in the coordinate system (X^A) have been denoted by $\lambda^A(\tau)$; the components of the connection symbols of H and H(t) have been denoted, of course, by $\Gamma^A_{BC}(X^D)$ and $\Gamma^A_{BC}(X^D, t)$ respectively. Now suppose that $\underset{\sim}{v}(\tau)$ is a vector field on λ which is obtained via parallel transport (from the vector $\underset{\sim}{v}(0)$ in the tangent space $B_{\lambda(0)}$) so that $\underset{\sim}{v}(\tau) = \rho(\tau)\underset{\sim}{v}(0)$ is a solution of (IV-44); then the vector field $\underset{\sim}{u}(\tau) = \underset{\sim}{\alpha}(\lambda(\tau), t)\underset{\sim}{v}(\tau)$ is a solution of (IV-45) because, by (IV-43),

$$\begin{aligned} \underset{\sim}{u}(\tau) &= \underset{\sim}{\alpha}(\lambda(\tau), t)\circ\rho(\tau)\underset{\sim}{v}(0) \qquad \text{(IV-46)} \\ &= \rho^t(\tau) \circ \underset{\sim}{\alpha}(\lambda(0), t)\underset{\sim}{v}(0) \\ &= \rho^t(\tau)\underset{\sim}{u}(0) \end{aligned}$$

Conversely, if $\underset{\sim}{u}(\tau) = \rho^t(\tau)\underset{\sim}{u}(0)$ satisfies (IV-45) then $\underset{\sim}{v}(\tau) = \underset{\sim}{\alpha}(\lambda(\tau), t)^{-1}\underset{\sim}{u}(\tau)$ is a solution of (IV-44). Therefore, if we denote the components of $\underset{\sim}{\alpha}$ relative to (X^A) by $\underset{\sim}{\alpha}^A_B(X^D, t)$, then

$$u^A(\tau) = \alpha^A_B(\lambda^D(\tau), t)v^B(\tau) \qquad \text{(IV-47)}$$

will be a solution of (IV-45) whenever $v^A(\tau)$ is a solution of (IV-44). Assuming that $v^A(\tau)$ <u>is</u> a solution of (IV-44) we now substitute (IV-47) into (IV-45) and obtain

$$\alpha^E_A[\frac{dv^A}{d\tau} + \alpha^{-1A}_{\ \ F}(\frac{\partial\alpha^F_B}{\partial X^C} + \Gamma^F_{DC}(\cdot,t)\alpha^D_B)v^B\frac{d\lambda^C}{d\tau} = 0 \qquad \text{(IV-48)}$$

If we now compare (IV-44) with (IV-48) we easily get

$$\Gamma^{A}_{BC}(\cdot) = \alpha^{-1A}_{F}(\cdot,t)\left(\frac{\partial\alpha^{F}_{B}(\cdot,t)}{\partial X^{C}} + \Gamma^{F}_{DC}(\cdot,t)\alpha^{D}_{B}(\cdot,t)\right) \qquad \text{(IV-49)}$$

which we may solve for $\Gamma^{A}_{BC}(\cdot,t)$ so as to obtain

<u>Theorem</u> IV-5. The connection symbols of an elastic material connection H on $\mathcal{B}$ and its associated anelastic material connection H(t) are related via

$$\Gamma^{A}_{BC}(\cdot,t) = \alpha^{A}_{F}(\cdot,t)\left(\frac{\partial\alpha^{-1F}_{B}(\cdot,t)}{\partial X^{C}} + \Gamma^{F}_{DC}(\cdot)\alpha^{-1D}_{B}(\cdot,t)\right) \qquad \text{(IV-50)}$$

$$= \alpha^{-1D}_{B}(\cdot,t)\left(\Gamma^{F}_{DC}(\cdot)\alpha^{A}_{F}(\cdot,t) - \frac{\partial\alpha^{A}_{D}(\cdot,t)}{\partial X^{C}}\right)$$

<u>Remark</u> Not only do (IV-49) and (IV-50) determine H(t) in terms of H and vice versa (relative to the anelastic transformation function $\underset{\sim}{\alpha}(\cdot,t)$) but in view of the intial condition for $\underset{\sim}{\alpha}$ we have $\Gamma^{A}_{BC}(\cdot,t) \to \Gamma^{A}_{BC}(\cdot)$ as $t\to-\infty$. The one-to-one correspondencd between H and H(t), which is given by (IV-49) and (IV-50) is not, of course, invariant under a change of the anelastic transformation function $\underset{\sim}{\alpha}(\cdot,t)$.

Now, it is possible to verify directly that the functions $\Gamma^{A}_{BC}(\cdot,t)$, which are defined via (IV-50), are the contion symbols of an anelastic material connection on $\mathcal{B}$ iff the functions $\Gamma^{A}_{BC}(\cdot)$ are the connection symbols of an elastic material connection on $\mathcal{B}$. To do this we select reference charts $(\mathcal{U},\underset{\sim}{r})$ and $(\mathcal{U},\underset{\sim}{r}(\cdot,t))$ in $\underset{\sim}{\Phi}$ and $\underset{\sim}{\Phi}(t)$, respectively, so that

$$\underset{\sim}{r}(\cdot,t) = \underset{\sim}{r}(\cdot) \circ \underset{\sim}{\alpha}(\cdot,t)^{-1}$$

where $\underset{\sim}{\alpha}(\cdot,t)$ is a smooth field of anelastic transformation functions defined on $\mathcal{B}$. If $\underset{\sim}{F}(\cdot) \equiv \kappa_* \circ \underset{\sim}{r}^{-1}$ is the deformation gradient from $\underset{\sim}{r}$ to κ_* (where $\kappa\colon \mathcal{B} \to R^3$) then the deformation gradient $\underset{\sim}{F}(\cdot,t)$ from $\underset{\sim}{r}(\cdot,t)$ to κ_* is given by

$$\underset{\sim}{F}(\cdot,t) = \underset{\sim}{F}(\cdot) \circ \underset{\sim}{r}(t,\cdot) \circ \underset{\sim}{\alpha}(\cdot,t) \circ \underset{\sim}{r}(\cdot,t)^{-1} \tag{IV-51}$$

or, in component form (relative to the coordinate system (X^A) which is induced on $\mathcal{B}$ by κ)

$$F^A_B(\cdot,t) = F^C_B(\cdot)\alpha^A_C(\cdot,t) \tag{IV-52}$$

Using (IV-52) and (IV-50) we now compute that

$$F^{-1A}_{C}(\cdot,t)\left(\frac{\partial F^C_B(\cdot,t)}{\partial X^D} + \Gamma^C_{ED}(\cdot,t)F^E_B(\cdot,t)\right) \tag{IV-53}$$

$$= F^{-1A}_{C}(\cdot)\left(\frac{\partial F^C_B(\cdot)}{\partial X^D} + \Gamma^C_{ED}(\cdot)F^E_B(\cdot)\right)$$

so that the matrices

$$[F^{-1A}_{C}\left(\frac{\partial F^C_B(\cdot)}{\partial X^D} + \Gamma^C_{ED}(\cdot)F^E_B(\cdot)\right),\ D = 1,2,3]$$

are contained in g_Φ iff the matrices

$$[F_C^{-1A}(\cdot,t)(\frac{\partial F_B^C(\cdot,t)}{\partial X^D} + \Gamma_{ED}^C(\cdot,t)F_B^E(\cdot,t)),\ D = 1,2,3]$$

are contained in $\underset{\sim}{g}_{\Phi}(t)(\equiv \underset{\sim}{g}_{\Phi}$; note that $\underset{\sim}{G}_{\Phi} = \underset{\sim}{G}_{\Phi(t)}$, $\forall t\varepsilon R$, by virtue if Theorem IV-4). But this last condition, relative to the Lie algebra $\underset{\sim}{g}_{\Phi(t)}$ of $\underset{\sim}{G}_{\Phi(t)}$, is precisely the one which guarantees that the functions $\Gamma_{BC}^A(\cdot,t)$ are the connection symbols of an anelastic material connection on $\mathcal{B}$.

Now let $\underset{\sim}{v}(\cdot)$ be any smooth vector field defined on $\mathcal{B}$. A direct computation, using (IV-50) and the definition of covariant differentiation,shows that the covariant derivative of $\underset{\sim}{v}(\cdot)$, relative to H, is related to that of the vector field $\underset{\sim}{\alpha}(\cdot,t)\underset{\sim}{v}(\cdot)$, relative to H(t), via

$$\alpha_B^A(\cdot,t)(\frac{\partial v^B(\cdot)}{\partial X^D} + \Gamma_{CD}^B(\cdot)v^C(\cdot)) \qquad \text{(IV-54)}$$

$$= \frac{\partial}{\partial X^D}(\alpha_B^A(\cdot,t)v^B(\cdot)) + \Gamma_{CD}^A(\cdot,t)\alpha_B^C(\cdot,t)v^B(\cdot)$$

or $\alpha_B^A v^B{}_{,D} = (\alpha_B^A v^B)_{|D}$, if we hold t fixed and denote covariant differentiation with respect to H and H(t) by "," and "|", respectively; an equivalent form of (IV-54) would be $\alpha_B^{-1A} v^B{}_{|D} = (\alpha_B^{-1A} v^B)_{,D}$

<u>Exercise</u> 25 Verify the relations given above for covariant differentiation of the contravariant vector field $\underset{\sim}{v}(\cdot)$. Show that the analogous results for a covariant vector

field $\underset{\sim}{w}(\cdot)$ are $\alpha^{-1B}_{A} w_{B,D} = (\alpha^{-1B}_{A} w_B)_{|D}$ and

$\alpha^B_A w_{B|D} = (\alpha^B_A w_B)_{,D}$. Verify that for an arbitrary tensor

field $\underset{\sim}{V}$ with components V^{AB}_{C} the appropriate formulas are

$$\alpha^A_P \alpha^{-1Q}_{C} \alpha^B_R V^{PR}_{Q,D} = (\alpha^A_P \alpha^{-1Q}_{C} \alpha^B_R V^{PR}_{Q})_{|D} \qquad \text{(IV-55)}$$

$$\alpha^{-1A}_{P} \alpha^Q_C \alpha^{-1B}_{R} V^{PR}_{Q|D} = (\alpha^{-1A}_{P} \alpha^Q_C \alpha^{-1B}_{R} V^{PR}_{Q})_{,D}$$

By using (IV-54) and its variant, it is a simple matter to prove the following theorem, which relates the torsion tensors of H and H(t):

<u>Theorem</u> IV-6 Let $\underset{\sim}{\theta}$ and $\underset{\sim}{\theta}(t)$ denote, respectively, the torsion tensors of H and H(t); then

$$\theta^A_{BC}(\cdot,t) = \theta^A_{BC}(\cdot) + \alpha^D_B \alpha^{-1A}_{D|C} - \alpha^D_C \alpha^{-1A}_{D|B} \qquad \text{(IV-56)}$$

$$= \theta^A_{BC}(\cdot) - \alpha^{-1D}_{B} \alpha^A_{D,C} + \alpha^{-1D}_{C} \alpha^A_{D,C}$$

<u>Proof</u> In terms of the covariant differention operation with respect to H and H(t), we may rewrite the two equations appearing in (IV-50) in the form

$$\Gamma^A_{BC}(\cdot,t) = \Gamma^A_{BC}(\cdot) - \alpha^{-1D}_{B} \alpha^A_{D,C} \qquad \text{(IV-57)}$$

$$= \Gamma^A_{BC}(\cdot) + \alpha^D_B \, \alpha^{-1A}_{D\ C}$$

The desired results, i.e. (IV-56), now follow directly from the definition $\theta^A_{BC} \equiv \Gamma^A_{BC} - \Gamma^A_{CB}$. We note in passing that $\underset{\sim}{\theta}(\cdot,t)$ depends smoothly on t in each motion $\phi(t)$ of $\mathcal{B}$ and that we also have

$$\underset{\sim}{\theta}(\cdot,t) \to \underset{\sim}{\theta}(\cdot), \text{ as } t\to-\infty$$

<u>Exercise</u> 26 Prove Theorem IV-6 by first applying the results of the last exercise to the gradient of an arbitrary smooth function f(•) so as to obtain

$$\alpha^B_A f_{|BD} = \alpha^B_A f_{,B} + \alpha^B_A f_{,BD} \tag{IV-58}$$

$$\alpha^{-1B}_{\ \ A} f_{,BD} = \alpha^{-1B}_{\ \ A|D} f_{|B} + \alpha^{-1B}_{\ \ A} f_{|BD}$$

and then applying the Ricci identity to (IV-58)

<u>Exercise</u> 27 Apply the general result (IV-55) for tensor fields to the covariant derivative of an arbitrary smooth vector field $\underset{\sim}{v}(\cdot)$ so as to obtain the result

$$v^B_{\ ,DE} = \alpha^C_D \alpha^{-1G}_{\ \ C|E} v^B_{\ ,G} + \alpha^{-1B}_{\ \ A} (\alpha^A_B v^G)_{|DE} \tag{IV-59}$$

Using the result of Exercise 27, above, we can now state following natural companion to Theorem IV-6, i.e.,

<u>Theorem</u> IV-7 Let $\underset{\sim}{\Omega}(\cdot)$ and $\underset{\sim}{\Omega}(\cdot,t)$ denote, respectively, the curvature tensors of H and H(t); then

$$\Omega^A_{BCD}(\cdot,t) = \alpha^A_E\, \alpha^{-1F}_{\ \ B}\, \Omega^E_{FCD}(\cdot) \tag{IV-60}$$

where the components of $\underset{\sim}{\Omega}$ are given by

$$\Omega^A_{BCD} = \frac{\partial\Gamma^A_{BD}}{\partial X^C} - \frac{\partial\Gamma^A_{BC}}{\partial X^D} + \Gamma^A_{EC}\Gamma^E_{BD} - \Gamma^A_{ED}\Gamma^E_{BC}$$

<u>Proof</u> the proof may be obtained via a direct computation based on the above formula for Ω^A_{BCD} and (IV-57) or by applying the Ricci identity to (IV-59); the former procedure gets quite involved and is not recommended.

Two corollories follow from the last two theorems. First of all, by virtue of (IV-56) it should be clear that even if $\theta^A_{BC}(\cdot) \equiv 0$ it does not follow that $\theta^A_{BC}(\cdot,t) \equiv 0$; therefore, if H is torsion-free then, in general, H(t) is not torsion-free. On the other hand, by virtue of (IV-60), $\Omega^A_{BCD}(\cdot,t) \equiv 0$ iff $\Omega^A_B(\cdot) \equiv 0$, so that H(t) is curvature-free iff H is; this can also be determined independent of the relation (IV-60) between $\underset{\sim}{\Omega}$ and $\underset{\sim}{\Omega}(\cdot,t)$, i.e., we need only note that $\underset{\sim}{\Omega} = 0$ is the necessary and sufficient condition for the local existence of solutions to the system $\underset{\sim}{v}^A_{,B} = 0$ which satisfy an initial condition of the form $\underset{\sim}{v}(p) = \underset{\sim}{v}_o$. However, we have already shown (i.e. (IV-46)) that $\underset{\sim}{v}$ is a solution of $v^A_{,B} = 0$ iff $\underset{\sim}{u} = \underset{\sim}{\alpha}\underset{\sim}{v}$ is a solution of $u^A_{|B} = 0$. It follows immediately that $\underset{\sim}{\Omega}(\cdot) \equiv 0$ iff $\underset{\sim}{\Omega}(\cdot,t) = \underset{\sim}{0}$ for some (and hence all) $t\varepsilon R$. As a further consequence of (IV-50) we also have

<u>Theorem</u> IV-7 $\Gamma^A_{AB}(\cdot,t) = \Gamma^A_{AB}(\cdot)$, $\forall t\varepsilon R$

Exercise 28. Prove Theorem IV-7 by establishing the formula

$$\Gamma^{A}_{AB}(\cdot,t) - \Gamma^{A}_{AB}(\cdot) = -\frac{\partial}{\partial X^{B}}(\log \det[\alpha^{C}_{D}(\cdot,t)])$$

and then using the fact that $\underset{\sim}{\alpha}(\cdot,t)$ is isochoric, i.e., $\det[\alpha^{A}_{B}(\cdot,t)] = 1$, $\forall t \varepsilon R$.

7. Anelastic Solid Bodies; Dislocation Motions

Let $\mathcal{B}$ be a smooth materially uniform anelastic solid body; the results of §6, Chapter II lead one to believe that some very interesting results may be obtained for two special kinds of anelastic solids, namely, solid crystals and isotropic solids and, indeed, such is the case. To begin with suppose that $\mathcal{B}$ is an anelastic solid crystal body and that $p \varepsilon \mathcal{B}$; then the anelastic transformation function $\underset{\sim}{\alpha}(p,t)$ must be unique in each motion $\phi(t)$ of $\mathcal{B}$ since the identity component $\hat{g}(p)$ of the symmetry group $g(p)$ is the trivial group consisting soley of the map $id_{\mathcal{B}_p}$ (Recall that if $\underset{\sim}{\alpha}(p,t)$ is any anelastic transformation function for p in the motion $\phi(t)$ then all anelastic transformation functions for p in the motion $\phi(t)$ are of the form $\underset{\sim}{\alpha}(p,t) \circ \underset{\sim}{\xi}(t)$ where $\underset{\sim}{\xi}(t) \ \varepsilon \ g(p)$, $\forall t \varepsilon R$, and $\underset{\sim}{\xi}(t) \to id_{\mathcal{B}_p}$ as $t \to -\infty$). Also, the elastic material connection H on $\mathcal{B}$ must be unique (where the connection symbols are, in fact, determined via (II-18)) so that the associated anelastic material connection H(t) is then uniquely determined by (IV-50); of course, both the elastic and anelastic material connections are curvature-free. The connection symbols of the unique anelastic material connection on an anelastic solid crystal body can also be determined via (IV-53) and the fact that $\underset{\sim}{g}_{\Phi(t)} = \{\underset{\sim}{0}\}$ so that for D = 1,2,3

$$[F^{-1A}_{\ C}(\cdot,t)(\frac{\partial F^C_B(\cdot,t)}{\partial X^D} + \Gamma^C_{ED}(\cdot,t)F^E_B(\cdot,t)]$$

is the zero matrix (i.e. see the argument on page IV-38. Because both H and H(t) are unique and curvature-free we may use the torsion tensors $\underset{\sim}{\theta}(\cdot)$ and $\underset{\sim}{\theta}(\cdot,t)$ to characterize, respectively, the elastic and anelastic inhomogeneity on an anelastic solid crystal body $\mathcal{B}$; we are led, therefore, to

Definition IV-5 Let $\underset{\sim}{\theta}(\cdot,t)$ be the torsion tensor associated with the unique anelastic material connection H(t) on a smooth materially uniform anelastic solid crystal body $\mathcal{B}$. Then we say that there exists a dislocation motion in $\mathcal{B}$ if $\frac{\partial \underset{\sim}{\theta}}{\partial t}(\cdot,t) \neq \underset{\sim}{0}$.

Having dispensed with the case where $\mathcal{B}$ is an anelastic solid crystal we now want to look briefly at the special case in which $\mathcal{B}$ is an anelastic isotropic solid body. As we have already indicated in §3, at each time t in a motion $\phi(t)\colon \mathcal{B} \to R^3$, we may define an intrinsic elastic metric m and an intrinsic anelastic metric m^t on $\mathcal{B}_p$ where p is any anelastic solid point; for the special case where p is an isotropic anelastic solid point we may choose an undistorted elastic reference atlas Φ and its corresponding anelastic reference atlas $\underset{\sim}{\Phi}(t)$ and then set

$$m(\underset{\sim}{u},\underset{\sim}{v}) \equiv \underset{\sim}{r}(p)\underset{\sim}{u} \cdot r(p)\underset{\sim}{v}, \ \forall \underset{\sim}{u},\underset{\sim}{v} \varepsilon \mathcal{B}_p \tag{IV-61}$$

and

$$m^t(\underset{\sim}{u},\underset{\sim}{v}) = \underset{\sim}{r}(p,t)\underset{\sim}{u} \cdot \underset{\sim}{r}(p,t)\underset{\sim}{v}, \ \forall \underset{\sim}{u},\underset{\sim}{v} \varepsilon \mathcal{B}_p \tag{IV-62}$$

where $(\mathcal{U},\underset{\sim}{r})$ and $(\mathcal{U},\underset{\sim}{r}(\cdot,t))$ are any reference charts contained in $\underset{\sim}{\Phi}$ and $\underset{\sim}{\Phi}(t)$, respectively, such that $p\varepsilon\mathcal{U}$.

Remark Note that since $\underset{\sim}{r}$ and $\underset{\sim}{r}(\cdot,t)$ are related via

$$\underset{\sim}{r}(\cdot,t) = \underset{\sim}{r}(\cdot) \circ \underset{\sim}{\alpha}(\cdot,t)^{-1}$$

on $\mathcal{U}$ it follows at once that m and m^t, as defined by (IV-61) and (IV-62), satisy (IV-16). Also, both m and m^t are smooth functions as the reference maps $\underset{\sim}{r}(\cdot)$ and $\underset{\sim}{r}(\cdot,t)$ are smooth. The following result, which we carry over directly from [19], is a consequence of Theorems II-4 and II-5 of Chapter II:

Theorem IV-8 The Riemannian connections H_R and $H_R(t)$ associated with the metrics m and m^t are the unique torsion-free elastic material connection and anelastic material connection on a smooth materially uniform isotropic anelastic solid body $\mathcal{B}$; the body $\mathcal{B}$ is locally elastically-homogeneous iff H_R is curvature-free and $\mathcal{B}$ is locally anelastically-homogeneous iff $H_R(t)$ is curvature-free.

From the statement of the theorem above it should be obvious that, just as we used the torsion tensors associated with the unique elastic and anelastic material connections on an anelastic solid crystal body to characterize the local elastic and anelastic inhomogeneities, we may use the curvature tensors of the Riemannian connections $H_{\mathcal{R}}$ and $H_{\mathcal{R}}(t)$ to characterize the distribution of elastic and anelastic inhomogeneity on an isotropic anelastic solid body $\mathcal{B}$. Unfortunately, we may not avail ourselves of equations (IV-50) when we seek the relationship between the connection symbols of $H_{\mathcal{R}}$ and those of $H_{\mathcal{R}}(t)$. In fact, $\underset{\sim}{\Phi}$, $\underset{\sim}{\Phi}(t)$, m, m^t, and $\underset{\sim}{\alpha}(\cdot,t)$ are not all unique in a given motion $\phi(t)$: $\mathcal{B} \to R^3$ while $H_{\mathcal{R}}$ and $H_{\mathcal{R}}(t)$ are unique and independent of the choice of these quantities; we therefore lose the isomorphism between elastic and anelastic material connections which is given by (IV-43).

<u>Exercise</u> 29 Verify the statement above by showing that if m is Euclidean while m^t is not then there can exist no anelastic transformation function $\underset{\sim}{\alpha}(\cdot,t)$ relative to which $H_{\mathcal{R}}$ is isomorphic to $H_{\mathcal{R}}(t)$ via (IV-43).

We now want to determine the relationship between the connection symbols of $H_{\mathcal{R}}$ and those of $H_{\mathcal{R}}(t)$. Once again we choose charts $(\mathcal{U},\underset{\sim}{r})$ and $(\mathcal{U},\underset{\sim}{r}(\cdot,t))$ in $\underset{\sim}{\Phi}$ and $\underset{\sim}{\Phi}(t)$, respectively, (with $\underset{\sim}{\Phi}$ undistorted) and set

$$\underset{\sim}{F}(\cdot) = \kappa_* \circ \underset{\sim}{r}^{-1}$$

$$\underset{\sim}{F}(\cdot,t) = \kappa_* \circ \underset{\sim}{r}(\cdot,t)^{-1}$$

where κ: $\mathcal{B} \to R^3$ induces coordinates (X^A) on $\mathcal{B}$; the relation ship between $\underset{\sim}{F}(\cdot)$ and $\underset{\sim}{F}(\cdot,t)$ is given by (IV-52). Now, the connection symbols of H_R are, of course, the Christoffel symbols of m, i.e.,

$$\left\{{A \atop BC}\right\} = \tfrac{1}{2}m^{AD}\left\{\frac{\partial m_{BD}}{\partial X^C} + \frac{\partial m_{CD}}{\partial X^B} - \frac{\partial m_{BC}}{\partial X^D}\right\} \tag{IV-63}$$

where m^{AB} and m_{AB} are the contravariant and covariant components of m in the reference configuration κ. Similarly, the connection symbols of $H_R(t)$ are the Christoffel symbols of m^t, i.e.,

$$\left\{{A \atop BC}\right\}(\cdot,t) = \tfrac{1}{2}m^{AD}(\cdot,t)\left(\frac{\partial m_{BD}(\cdot,t)}{\partial X^C} + \frac{\partial m_{CD}(\cdot,t)}{\partial X^B} - \frac{\partial m_{BC}(\cdot,t)}{\partial X^D}\right) \tag{IV-64}$$

It is a simple matter to show that (IV-61) and (IV-62) have the following covariant and contravariant component representations:

$$m^{AB} = \delta^{CD}F^A_C F^B_D$$

$$m_{AB} = \delta_{CD}F^{-1C}{}_A F^{-1D}{}_B$$

$$m^{AB}(\cdot,t) = \delta^{CD}F^A_C(\cdot,t)F^B_D(\cdot,t)$$

$$m_{AB}(\cdot,t) = \delta_{CD}F^{-1C}{}_A(\cdot,t)F^{-1D}{}_B(\cdot,t)$$

which are related to each other via the transformation rules (relative to some choice of a specific anelastic transformation function $\underset{\sim}{\alpha}(\cdot,t)$)

$$m^{AB}(\cdot,t) = \alpha^A_C(\cdot,t)\alpha^B_D(\cdot,t)m^{CD}(\cdot)$$
$$m_{AB}(\cdot,t) = \alpha^{-1C}_{\ \ A}(\cdot,t)\alpha^{-1D}_{\ \ B}(\cdot,t)m_{CD}(\cdot) \qquad \text{(IV-65)}$$

Note that (IV-65) and the initial condition on $\underset{\sim}{\alpha}(\cdot,t)$ implies that $m^t \to m$ as $t\to-\infty$; it then follows from (IV-63) and (IV-64) that $H_R(t) \to H$ as $t\to-\infty$. Now, we can obtain the desired relation between $\{^{\ A}_{BC}\}$ and $\{^{\ A}_{BC}\}(\cdot,t)$ if we substitute from (IV-65) into (IV-64) and then make use of (IV-63) but this approach becomes extremely involved; an alternate approach may be based on the properties of the covariant derivative and leads to the following

<u>Theorem</u> IV-9 The Christoffel symbols of the Riemannian connections H_R and $H_R(t)$, associated with the metrics m and m^t, respectively, are related by

$$\{^{\ A}_{BC}\}(\cdot,t) = \{^{\ A}_{BC}\} + \tfrac{1}{2}\alpha^A_H\alpha^D_J[(\alpha^{-1E}_{\ \ B}\alpha^{-1F}_{\ \ D})_{,C} \qquad \text{(IV-66)}$$
$$+ (\alpha^{-1E}_{\ \ C}\alpha^{-1F}_{\ \ D})_{,B} - (\alpha^{-1E}_{\ \ B}\alpha^{-1F}_{\ \ C})_{,D}]m^{HJ}m_{EF}$$

where $\alpha^A_B(\cdot,t)$ are the components of some anelastic transformation function for $\mathcal{B}$ in some motion $\phi(t)$.

<u>Proof</u>: Following Wang & Bloom [19] we define a tensor field

$\underset{\sim}{\Delta}(\cdot,t)$ on $\kappa(\mathcal{B})$ via

$$\Delta^A_{BC}(\cdot,t) \equiv \{^{\,A}_{BC}\}(\cdot,t) - \{^{\,A}_{BC}\}(\cdot) \tag{IV-67}$$

Clearly, $\Delta^A_{BC} = \Delta^A_{CB}$. Once again we hold t fixed and denote covariant differentiation with respect to H_R and $H_R(t)$ by " , " and "|", respectively. Then if $\underset{\sim}{v}$ is a vector field on $\mathcal{B}$ and $\underset{\sim}{w}$ is a covector field simple computations produce

$$v^A_{\ |B} - v^A_{\ ,B} = \Delta^A_{CB} v^C \tag{IV-68}$$

and

$$w_{A|B} - w_{A,B} = -\Delta^C_{AB} w_C \tag{IV-69}$$

These results will not be needed until we come to the next theorem; however, a result, in the same direction as (IV-68) and (IV-69), which is needed here is

$$\begin{aligned} m_{AB|C}(\cdot,t) - m_{AB,C}(\cdot,t) &= -m_{DB}(\cdot,t)\Delta^D_{AC} \\ &\quad - m_{DA}(\cdot,t)\Delta^D_{BC} \end{aligned} \tag{IV-70}$$

which is equivalent to

$$\begin{aligned} m_{DB}(\cdot,t)\Delta^D_{AC}(\cdot,t) &+ m_{DA}(\cdot,t)\Delta^D_{BC} \\ &= m_{AB,C}(\cdot,t) \end{aligned} \tag{IV-71}$$

since m^t is horizontal with respect to $H_R(t)$(i.e. $m_{AB|C}(\cdot,t) \equiv 0$

because m^t, being an intrinsic anelastic inner product, must be invariant under all induced transformations via anelastic material isomorphisms). If we now take the co-variant derivative with respect to H_R on both sides of (IV-65_2) and use the fact that m is horizontal with respect to H_R, i.e., that $m_{AB,C}(\cdot) = 0$, we get

$$m_{AB,C}(\cdot,t) = (\alpha^{-1E}_{\ A}\alpha^{-1F}_{\ B})_{,C}m_{EF}(\cdot) \tag{IV-72}$$

Combining (IV-71) and (IV-72) we have

$$\begin{aligned} m_{DB}(\cdot,t)\Delta^D_{AC} + m_{DA}(\cdot,t)\Delta^D_{BC} \\ = (\alpha^{-1E}_{\ A}\alpha^{-1F}_{\ B})_{,C}m_{EF}(\cdot) \end{aligned} \tag{IV-73}$$

and so if we use the symmetry condition on $\underset{\sim}{\Delta}$ and solve for Δ^A_{BC} we find

$$\begin{aligned} \Delta^A_{BC}(\cdot,t) &= \tfrac{1}{2}m^{AD}(\cdot,t)[(\alpha^{-1E}_{\ B}\alpha^{-1F}_{\ D})_{,C} \\ &\quad + (\alpha^{-1E}_{\ C}\alpha^{-1F}_{\ D})_{,B} - (\alpha^{-1E}_{\ B}\,{}^{-1F}_{\ C})_{,D}]m_{EF}(\cdot) \\ &= \tfrac{1}{2}\alpha^A_H\alpha^D_J[(\alpha^{-1E}_{\ B}\alpha^{-1F}_{\ D})_{,C} \\ &\quad + (\alpha^{-1E}_{\ C}\alpha^{-1F}_{\ D})_{,B} - (\alpha^{-1E}_{\ B}\alpha^{-1F}_{\ C})_{,D}]m^{HJ}(\cdot)m_{EF}(\cdot) \end{aligned} \tag{IV-74}$$

from which the required result follows. Q. E. D.

<u>Exercise</u> 30 Verify that the right-hand side of (IV-74_2) is

independent of the choice of the intrinsic elastic metric m and the anelastic transformation function $\underset{\sim}{\alpha}(\cdot,t)$.

Now, as the distribution of elastic and anelastic inhomogeneities on an isotropic anelastic solid body may be characterized by the respective curvature tensors of $H_{\mathcal{R}}$ and $H_{\mathcal{R}}(t)$ the following theorem, which delineates the precise relationship between these two curvature tensors, is of considerable interest:

<u>Theorem</u> IV-10 The curvature tensors $\underset{\sim}{\Omega}(\cdot,t)$ and $\underset{\sim}{\Omega}$ which are associated, respectively, with $H_{\mathcal{R}}(t)$ and $H_{\mathcal{R}}$, are related by

$$\Omega^{A}_{ECB}(\cdot,t) = \Omega^{A}_{ECB} + (\Delta^{A}_{EB,C} - \Delta^{A}_{EC,B} + \Delta^{A}_{DC}\Delta^{D}_{EB} - \Delta^{A}_{DB}\Delta^{D}_{EC}) \qquad \text{(IV-75)}$$

<u>Proof</u>: For the tensor field with components $v^{A}{}_{|B}$, where $\underset{\sim}{v}(\cdot)$ is a vector field on $\mathcal{B}$, we easily compute

$$v^{A}{}_{|BC} - v^{A}{}_{|B,c} = \Delta^{A}_{DC}v^{D}{}_{|B} - \Delta^{D}_{BC}v^{A}{}_{|D} \qquad \text{(IV-76)}$$

If we now take the covariant derivative of (IV-68) with respect to $H_{\mathcal{R}}$ we obtain

$$v^{A}{}_{|B,C} - v^{A}{}_{,BC} = \Delta^{A}_{DB,C}v^{D} + \Delta^{A}_{DB}v^{D}{}_{,C} \qquad \text{(IV-77)}$$

and addition of (IV-76) and (IV-77) then produces

$$v^{A}{}_{|BC} - v^{A}{}_{,BC} = \Delta^{A}_{DB,C}v^{D} + \Delta^{A}_{DB}v^{D}{}_{,C} + \Delta^{A}_{DC}v^{D}{}_{|B} - \Delta^{D}_{BC}v^{A}{}_{|D} \qquad \text{(IV-78)}$$

The proof is now completed by using (IV-68) to rewrite (IV-78) in the form

$$v^A_{|BC} - v^A_{,BC} = \bar{\Delta}^A_{EBC} v^E \qquad \text{(IV-79)}$$

$$+ (\Delta^A_{DB} v^D_{,C} + \Delta^A_{DC} v^D_{,B} - \Delta^D_{BC} v^A_{,D})$$

where $\bar{\Delta}^A_{EBC} \equiv \Delta^A_{EB,C} + \Delta^A_{DC}\Delta^D_{EB} - \Delta^D_{BC}\Delta^A_{ED}$, and then taking the skew symmetric part of (IV-79), with respect to the indices B,C and using Ricci's identity.

Exercise 31 Follow the directions given above and verify the formula (IV-75).

Exercise 32 Show that $\{^{A}_{AB}\}(\cdot,t) = \{^{A}_{AB}\}(\cdot)$, $\forall t \varepsilon R$.
hint: from (IV-65) and the fact that $\underset{\sim}{\alpha}(\cdot,t)$ is isochoric, $t\varepsilon R$, if follows that

$$\det[m_{AB}(\cdot,t)] = \det[m_{AB}(\cdot)], \quad \forall t \varepsilon R$$

so that the result follows immediately if we can verify the formula

$$\frac{\partial}{\partial X^B}(\log\sqrt{\det[m_{CD}]}) = \{^{A}_{AB}\}$$

8. Equations of Motion for Anelastic Bodies

The derivation of the field equations of motion for smooth materially uniform anelastic bodies is a very simple matter once we have, at hand, the corresponding equations of motion for smooth materially uniform elastic bodies. On the other hand, to derive the equations of motion for anelastic bodies directly, we would begin with the Cauchy equations (II-24) and then select an anelastic reference chart $(\mathcal{U}, r(\cdot,t)) \in \underset{\sim}{\Phi}(t)$ and a global configuration $\kappa: \mathcal{B} \to R^3$. In place of (II-23) we would have

$$T^i_j(x) = S^i_j(F^p_q(x,t)) \qquad \text{(IV-80)}$$

as $\underset{\sim}{F}(\cdot,t)$ is the deformation gradient from $\underset{\sim}{r}(\cdot,t)$ to κ_*; the $F(\cdot)$ appearing in (II-23), is, of course, the deformation gradient from $\underset{\sim}{r}(\cdot)$ to κ_* where $(\mathcal{U},\underset{\sim}{r}(\cdot)) \in \underset{\sim}{\Phi}$, the elastic reference atlas which is associated with $\underset{\sim}{\Phi}(t)$. Note also that by virtue of Theorem IV-4, the response function $S^t_{r(\cdot,t)}(\underset{\sim}{F},\cdot)$ on $\mathcal{U}$ coincides with $S_{r(\cdot)}(\underset{\sim}{F},\cdot)$, for all $t \in R$ and all $F \in GL(3)$, so that the functions S^i_j appearing in (IV-80) are, essentially, the same as those which appear in II-23. To complete the derivation we need only follow the steps analogous to those used in proceeding from (II-24) through (II-29); the major difference is that in addition to replacing $\underset{\sim}{\Phi}$ by $\underset{\sim}{\Phi}(t)$ we would now replace the (elastic) material connection H employed in Chapter II by an anelastic material connection H(t). As it turns out, a direct derivation is not required for we can simply make the required changes, i.e., $\underset{\sim}{\Phi} \to \underset{\sim}{\Phi}(t)$, $\underset{\sim}{F}(\cdot) \to \underset{\sim}{F}(\cdot,t)$, and $H \to H(t)$, in the global field equations of motion (II-31); for the sake of convenience we rewrite these equations below in the form

$$\tilde{H}^{iA}_{jk} x^k_{A,B} X^B_j + \rho b^i = \rho \ddot{x}^i \qquad \text{(IV-81)}$$

where $\phi(t)\colon \mathcal{B} \to \mathcal{R}^3$ is represented by the deformation functions $x^i(X^A,t)$ and

$$x^i_A = \frac{\partial x^i(X^B,t)}{\partial X^A}, \quad X^A_i = \frac{\partial X^A(x^j,t)}{\partial x^i}$$

In (IV-81) $x^k_{A,B}$ denotes the covariant derivative of x^k_A relative to some elastic material connection H, i.e.,

$$x^k_{A,B} = \frac{\partial^2 x^k}{\partial X^A \partial X^B} - \Gamma^C_{AB} \frac{\partial x^k}{\partial X^C}$$

where the Γ^A_{BC} are the connection symbols of H; also $\tilde{H}^{iA}_{jk}$ is a global field defined by

$$\tilde{H}^{iA}_{jk} = H^{iB}_{jk}([x^l_G F^G_H]) F^A_B,$$

where $H^{iB}_{jk} = \partial S^i_j / \partial F^k_A$. If in (IV-81) we carry out the transformation from $\Gamma^A_{BC}(\cdot) \to \Gamma^A_{BC}(\cdot,t)$ then the covariant derivative of x^k_A with respect to H is replaced by the covariant derivative $x^k_{A|B}$ (of x^k_A with respect to H(t)). As we go from $\underset{\sim}{\Phi} \to \underset{\sim}{\Phi}(t)$ we also make the transformation

$$\underset{\sim}{F}(\cdot) \to \underset{\sim}{F}(\cdot,t) = \underset{\sim}{F}(\cdot) \circ \underset{\sim}{r}(\cdot,t) \circ \underset{\sim}{\alpha}(\cdot,t) \circ \underset{\sim}{r}(\cdot,t)^{-1}$$

or, in component form, $F^A_B(\cdot) \to F^A_B(\cdot,t) = \alpha^A_C(\cdot,t) F^C_B(\cdot)$; making these various changes in (IV-81) we get the global field equations of motion for a smooth materially uniform anelastic body, i.e.,

$$\hat{H}^{iA}_{jk}\, x^k_{A|B} X^B_j + \rho b^i = \rho \ddot{x}^i \tag{IV-82}$$

where

$$\hat{H}^{iA}_{jk} \equiv H^{iB}_{jk}([x^1_F \alpha^F_G F^G_H]) \alpha^A_C F^C_B \tag{IV-83}$$

Recall now (i.e., eq. (IV-51)) that $\alpha^A_C x^k_{A|B} = (\alpha^A_C x^k_A)_{,B}$ for $k = 1,2,3$. If we substitute this result into (IV-82) then we get

$$\bar{H}^{iA}_{jk}(x^k_D \alpha^D_A)_{,B}\, X^B_j + \rho b^i = \rho \ddot{x}^i \tag{IV-84}$$

where

$$\bar{H}^{iA}_{jk} = H^{iB}_{jk}([x^1_F \alpha^F_G F^G_H]) F^A_B \tag{IV-85}$$

A quick glance at (IV-84), (IV-85) shows that this system of equations is formally the same as those for a smooth materially uniform elastic body (i.e., (IV-81)) except that the variable x^1_G has been replaced throughout the system by the variable $x^1_F \alpha^F_G$.

<u>Exercise</u> 33 Derive the variant of the system of equations (II-33) which is applicable for a smooth materially uniform anelastic body $\mathcal{B}$.

Chapter V. Thermodynamics and Dislocation Motion

1. Introduction

In this final chapter we want to generalize the mechanical model of anelastic response in such a way as to allow for thermodynamical influences; our work here is based on an earlier paper of Wang [36], which dealt with the thermodynamics of inhomogeneous bodies, and on a recent paper by Bloom and Wang [20]. Three kinds of basic results may be found in the present chapter. First of all, we derive those restrictions on the constitutive functions of thermo-anelastic points and their associated flow rules which are implied by the well-known Clausius-Duhem inequality (alternatively, the "entropy inequality"). Secondly, we introduce the concept of symmetry isomorphism and use it to generate the respective thermoelastic and anelastic global structures and their associated geometric structures. Finally, we use the induced geometric structures to derive global thermodynamic field equations of balance for both (inhomogeneous) thermoelastic bodies and thermo-anelastic bodies with uniform symmetry. As the reader will soon discern, by replacing the previous concept of <u>material uniformity</u> with the concept of <u>symmetry uniformity</u>, we will be substituting for the assumption that the material points of our body have the same material response the weaker assumption that they have the same material symmetry.

2. The Concept of a Thermoelastic Point and the Entropy Inequality.

Let $\mathcal{B}$ be a material body and let $p \in \mathcal{B}$; we say that p is a thermoelastic point[14] if the stress at p, in any local configuration $\underset{\sim}{\nu}(p) \in \mathcal{D}_p$, as well as the internal energy function ε, the entropy η and the heat flux vector $\underset{\sim}{q}$ at $\underset{\sim}{\nu}(p)$ are determined by constitutive equations which are of the form

$$\underset{\sim}{T}_S = \underset{\sim}{E}_p(\underset{\sim}{\nu}(p), \theta(p), \underset{\sim}{g}(p)) \tag{V-1}$$

$$\varepsilon = e_p(\underset{\sim}{\nu}(p), \theta(p), \underset{\sim}{g}(p)) \tag{V-2}$$

$$\eta = h_p(\underset{\sim}{\nu}(p), \theta(p), \underset{\sim}{g}(p)) \tag{V-3}$$

$$\underset{\sim}{q} = \underset{\sim}{\ell}_p(\underset{\sim}{\nu}(p), \theta(p), \underset{\sim}{g}(p)) \tag{V-4}$$

where $\theta(p)$ is the local temperature at p and $\underset{\sim}{g}(p)$ is the local temperature gradient at p in the local configuration $\underset{\sim}{\nu}(p)$; in a local motion we would, of course, replace $\underset{\sim}{\nu}(p) \to \underset{\sim}{\nu}(p,t)$, $\theta(p) \to \theta(p,t)$, and $\underset{\sim}{g}(p) \to \underset{\sim}{g}(p,t)$. The p subscript on the constitutive functions in (V-1) through (V-4) indicates the explicit dependence of these functions on the point p; of the new functions appearing above e_p and h_p are scalar-valued for each $p \in \mathcal{B}$ while $\underset{\sim}{\ell}_p$ takes its values in R^3 for each $p \in \mathcal{B}$.

(14) see, also, example (ii) on page III-4.

As we have done in the previous chapters we may rewrite our constitutive equations by introducing a preferred local configuration $\underset{\sim}{r}(p)\varepsilon\mathcal{D}_p$ which we will call, as usual, a local reference configuration. With $\underset{\sim}{F} = \underset{\sim}{\nu}(p) \circ \underset{\sim}{r}(p)^{-1}$ as the local deformation from $\underset{\sim}{r}(p)$ to $\underset{\sim}{\nu}(p)$ we may rewrite (V-1) through (V-4) in the form

$$\underset{\sim}{T}_S = \underset{\sim}{S}_{\underset{\sim}{r}(p)}(\underset{\sim}{F},\theta(p),\ \underset{\sim}{g}(p),\ p) \qquad \text{(V-1')}$$

$$\varepsilon = \mathcal{E}_{\underset{\sim}{r}(p)}(\underset{\sim}{F},\theta(p),\ \underset{\sim}{g}(p),\ p) \qquad \text{(V-2')}$$

$$\eta = \mathcal{H}_{\underset{\sim}{r}(p)}(\underset{\sim}{F},\theta(p),\ \underset{\sim}{g}(p),\ p) \qquad \text{(V-3')}$$

$$\underset{\sim}{q} = \underset{\sim}{\mathcal{L}}_{\underset{\sim}{r}(p)}(\underset{\sim}{F},\theta(p),\ \underset{\sim}{g}(p),\ p) \qquad \text{(V-4')}$$

where

$$\underset{\sim}{S}_{\underset{\sim}{r}(p)}(\underset{\sim}{F},\theta(p),\ \underset{\sim}{g}(p),\ p) \equiv \underset{\sim}{E}_p(\underset{\sim}{F}\circ\underset{\sim}{r}(p),\ \theta(p),\ \underset{\sim}{g}(p))$$

for all $\underset{\sim}{F}\varepsilon GL(3)$ and the other functions appearing in (V-1') through (V-4') are defined in a similar manner; note that the dependence of the response functions on p has been shown explicitly by grouping p with the arguments of these functions.

When considering thermoelastic bodies (i.e., material bodies composed of thermoelastic points) it is not sufficient to restrict ones attention to mechanical balance laws such as the principle of balance of momentum, which yields, as a consequence (given sufficient smoothness conditions on the quantities involved) Cauchy's equations of motion II-24, or the principle of balance of moment of momentum, which yields as a consequence the symmetry of the Cauchy Stress tensor $\underset{\sim}{T}_S$. We must also consider the so-called first and second laws of thermodynamics which are also termed the principal of energy balance and the entropy inequality or Clausius-Duhem inequality; a combination of these latter two principles can be shown (i.e., see [5]) to lead to the <u>reduced entropy inequality</u>[15]

$$\rho(\dot{\psi}+\eta\dot{\theta}) - \mathrm{tr}(\underset{\sim}{T}_S \text{ grad } \underset{\sim}{v}) + \frac{q \cdot g}{\theta} \leq 0 \qquad \text{(V-5)}$$

It was demonstrated by Coleman and Noll [37] that the reduced entropy inequality (V-5) leads to the following restrictions on the constitutive functions appearing in (V-1') through (V-4') if these constitutive relations are to be consistent with the second law of thermodynamics[16]:

I. the vector function $\underset{\sim}{L}_{r(\underset{\sim}{p})}$ must satisfy the inequality

$$\underset{\sim}{L}_{r(\underset{\sim}{p})}(\underset{\sim}{F},\theta,\underset{\sim}{g},p) \cdot \underset{\sim}{g} \leq 0 \qquad \text{(V-6)}$$

(15) v represents the velocity in some motion $\phi(t)$ of $\mathcal{B}$ and $\psi \equiv \varepsilon - \eta\theta$ is called the free-energy function; ρ is the density in the configuration $\phi(t)$ at time t.

(16) for a detailed proof see [37].

for each $p \in \mathcal{B}$ and all $(\underset{\sim}{F}, \theta, \underset{\sim}{g})$.

II. the functions $\underset{\sim}{S}_{r(p)}$, $\mathcal{E}_{\underset{\sim}{r}(p)}$ and $\mathcal{H}_{\underset{\sim}{r}(p)}$ are all independent of the variable $\underset{\sim}{g}(p)$ for each $p \in \mathcal{B}$.

III. there exists a function $\psi_{\underset{\sim}{r}(p)}(F, \theta, p)$ called the free-energy (see footnote (15) below) such that for each $p \in \mathcal{B}$

$$S_{\underset{\sim}{r}(p)}(\underset{\sim}{F}, \theta, p) = \rho \underset{\sim}{F} \frac{\partial}{\partial \underset{\sim}{F}} \psi_{\underset{\sim}{r}(p)}(F, \theta, p)^{T} \tag{V-7}$$

$$\mathcal{E}_{\underset{\sim}{r}(p)}(\underset{\sim}{F}, \theta, p) = \psi_{\underset{\sim}{r}(p)}(\underset{\sim}{F}, \theta, p) + \theta \frac{\partial}{\partial \theta} \psi_{\underset{\sim}{r}(p)}(\underset{\sim}{F}, \theta, p) \tag{V-8}$$

and

$$\mathcal{H}_{\underset{\sim}{r}(p)}(\underset{\sim}{F}, \theta, p) = \frac{\partial}{\partial \theta} \psi_{\underset{\sim}{r}(p)}(\underset{\sim}{F}, \theta, p) \tag{V-9}$$

The symmetry group $G^{*}_{\underset{\sim}{r}(p)}(p)$ is defined to be the set of all maps $\underset{\sim}{K} \in SL(3)$ such that for all $(\underset{\sim}{F}, \theta)$

$$\psi_{\underset{\sim}{r}(p)}(\underset{\sim}{F}\underset{\sim}{K}, \theta, p) = \psi_{\underset{\sim}{r}(p)}(\underset{\sim}{F}, \theta, p) \tag{V-10}$$

<u>Exercise</u> 34. Show that $G^{*}_{\underset{\sim}{r}(p)}(p)$ is contained in the symmetry groups of $\mathcal{E}_{\underset{\sim}{r}(p)}$ and $\mathcal{H}_{\underset{\sim}{r}(p)}$.

<u>Remark</u> It has been proven by Truesdell [38] that $\underset{\sim}{K} \in G_{\underset{\sim}{r}(p)}(p)$, the symmetry group of $S_{r(p)}$ iff

$$\psi_{\underset{\sim}{r}(p)}(\underset{\sim}{F}\underset{\sim}{K},\theta,p) = \psi_{\underset{\sim}{r}(p)}(\underset{\sim}{F},\theta,p) \tag{V-11}$$

$$+ \psi_{\underset{\sim}{r}(p)}(\underset{\sim}{K},\theta,p) - \psi_{\underset{\sim}{r}(p)}(\underset{\sim}{1},\theta,p)$$

so that, in general, $G_{\underset{\sim}{r}(p)} \neq G^*_{\underset{\sim}{r}(p)}(p)$.

We now make a basic assumption which will be crucial for the development of the material geometric structure theory for thermoelastic and thermo-anelastic material bodies with uniform symmetry, i.e., we will assume that the symmetry group of $L_{\underset{\sim}{r}(p)}$, for any $p\varepsilon B$, coincides with $G_{\underset{\sim}{r}(p)}(p)$, the symmetry group of $\underset{\sim}{S}_{\underset{\sim}{r}(p)}$ (we note here that the results of Coleman and Noll [37] do not relate, in any definite way, the symmetry group of $\underset{\sim}{L}_{\underset{\sim}{r}(p)}$ to those of either $\psi_{\underset{\sim}{r}(p)}$ or $\underset{\sim}{S}_{\underset{\sim}{r}(p)}$.) Under this assumption the symmetry groups of $\underset{\sim}{S}_{\underset{\sim}{r}(p)}$ and $\underset{\sim}{L}_{\underset{\sim}{r}(p)}$ will be closed Lie subgroups of SL(3) since the free energy function $\psi_{\underset{\sim}{r}(p)}$ is usually assumed to be a smooth function for each $p\varepsilon B$.

The transformation laws for the various symmetry groups introduced above closely resemble those which we have met in the previous chapters, i.e., if $\bar{r}(p) \varepsilon D_p$ is any other local reference configuration of $p\varepsilon B$ and we transform from $\underset{\sim}{r}(p) \to \underset{\sim}{\bar{r}}(p)$ then

$$G_{\bar{r}(p)}(p) = \underset{\sim}{L} G_{\underset{\sim}{r}(p)}(p)\underset{\sim}{L}^{-1} \tag{V-12}$$

where $\underset{\sim}{L} \equiv \underset{\sim}{\bar{r}}(p) \circ \underset{\sim}{\bar{r}}(p)^{-1}$ and

$$G^*_{\bar{r}(p)}(p) = LG^*_{r(p)}(p)\underset{\sim}{L}^{-1} \tag{V-13}$$

Both (V-12) and (V-13) may be proved by introducing the symmetry groups $\mathcal{g}(p)$ and $\mathcal{g}^*(p)$ which are defined via

$$\mathcal{g}(p) = \{\underset{\sim}{A} \varepsilon SL(\mathcal{B}_p) \mid \mathcal{E}_p(\underset{\sim}{\nu}(p) \circ \underset{\sim}{A}, \theta(p))$$

$$\underset{\sim}{\mathcal{E}}_p(\underset{\sim}{\nu}(p), \theta(p)) \; \forall \underset{\sim}{\nu}(p) \; \varepsilon \; \mathcal{D}_p \text{ (at each fixed}$$

value of $\theta(p))\}$

and

$$\mathcal{g}^*(p) = \{\underset{\sim}{A} \varepsilon SL(\mathcal{B}_p) \mid \psi(\underset{\sim}{\nu}(p) \circ \underset{\sim}{A}, \; \theta(p), p) =$$

$$\psi(\underset{\sim}{\nu}(p), \theta(p), p), \; \forall \underset{\sim}{\nu}(p) \; \varepsilon \; \mathcal{D}_p, \text{ at each } \theta(p)\}$$

where

$$\psi(\underset{\sim}{\nu}(p), \theta(p), p) \equiv \psi_{\underset{\sim}{r}(p)}(\underset{\sim}{F}, \theta, p), \; \underset{\sim}{F} \equiv \underset{\sim}{\nu}(p) \circ \underset{\sim}{r}(p)^{-1}.$$

The relations (V-12) and (V-13) now follow, respectively, from the results

$$G_{\underset{\sim}{r}(p)}(p) = \underset{\sim}{r}(p) \circ \mathcal{g}(p) \circ \underset{\sim}{r}(p)^{-1} \qquad \text{(V-14)}$$

$$G^*_{\underset{\sim}{r}(p)}(p) = \underset{\sim}{r}(p) \circ \mathcal{g}^*(p) \circ \underset{\sim}{r}(p)^{-1} \qquad \text{(V-15)}$$

which are a direct consequence of the definitions of the various symmetry groups involved and the relations which exist among their associated response functions.

3. Geometric Structures on Thermoelastic Bodies with Uniform Symmetry

In Chapters II-IV we have based the construction of

different kinds of geometric structures, for several kinds of material bodies, on various concepts of material uniformity; in this section we want to show how a geometric theory may be constructed for a thermoelastic body which exhibits <u>uniform symmetry</u>. In order to make this latter concept precise we first state

<u>Definition</u> V-1 Let p and q be any two points which belong to the thermoelastic body $\mathcal{B}$. Then a linear isomorphism $\underset{\sim}{s}(p,q)\colon \mathcal{B}_p \to \mathcal{B}_q$ is called a <u>symmetry isomorphism</u> of p and q if the symmetry groups $\mathcal{g}(p)$ and $\mathcal{g}(q)$ are related via

$$\mathcal{g}(q) = \underset{\sim}{s}(p,q) \circ \mathcal{g}(p) \circ \underset{\sim}{s}(p,q)^{-1} \qquad \text{(V-16)}$$

We may note, first of all, that it is an immediate consequence of (V-14) and the definition above that $\underset{\sim}{s}(p,q) \equiv \underset{\sim}{r}(q)^{-1} \circ \underset{\sim}{r}(p)$ defines a symmetry isomorphism of p and q whenever $\underset{\sim}{r}(p)$ and $\underset{\sim}{r}(q)$ are local (reference) configurations of p and q, respectively, which satisfy $G_{r(p)}(p) = G_{r(q)}(q)$. Conversely, if $\underset{\sim}{s}(p,q)$ is a symmetry isomorphism of p and q of the form $\underset{\sim}{s}(p,q) = \underset{\sim}{r}(q)^{-1} \circ \underset{\sim}{r}(p)$, for some $\underset{\sim}{r}(p) \in \mathcal{D}_p$, $\underset{\sim}{r}(q) \in \mathcal{D}_q$, then the symmetry groups of $\underset{\sim}{S}_{r(p)}(\underset{\sim}{F},\theta,p)$ and $\underset{\sim}{S}_{\underset{\sim}{r}(q)}(\underset{\sim}{F},\theta,q)$ must coincide, that is, $G_{\underset{\sim}{r}(p)}(p) = G_{\underset{\sim}{r}(q)}(q)$.

<u>Remarks</u> Recall that a material isomorphism of p and q (in the present situation) would be a linear isomorphism

$\underset{\sim}{I}(p,q)\colon \mathcal{B}_p \to \mathcal{B}_q$ which satisfies (i.e., see equation (II-3))

$$\underset{\sim}{\mathcal{E}}_q(\underset{\sim}{\nu}(q),\ \theta) = \mathcal{E}_p(\underset{\sim}{\nu}(q) \circ \underset{\sim}{I}(p,q),\ \theta) \qquad \text{(V-17)}$$

for all $\underset{\sim}{\nu}(q) \in \mathcal{D}_q$ and each fixed θ; we might even place a θ supscript on I, in this case,to indicate the fact that it will depend, in general, on θ. If p and q are materially isomorphic then we know (i.e., see equation (II-8)) that (V-16) is satisfied with $\underset{\sim}{s}(p,q)$ replaced by $\underset{\sim}{I}(p,q)$; therefore, every material isomorphism is, a priori, a symmetry isomorphism. Clearly, there is no reason why an arbitrary symmetry isomrophism should also be a material isomorphism. In other words, the fact that material points of $\mathcal{B}$ have the same material response implies that the relative symmetry groups of the body points are all the same, in the sense that the conjugate classes of the relative symmetry groups of different point of $\mathcal{B}$ coincide with one another; the converse of this statement is, in general, false. Naturally we say that $\mathcal{B}$ is a <u>thermoelastic body with symmetry uniformity</u> if for each pair $p,q \in \mathcal{B}$ there exists a linear (symmetry) isomorphism satisfying (V-16).

In order to the develop a geometric theory for thermoelastic bodies with uniform symmetry we introduce the concept of a <u>symmetry chart</u> $(\mathcal{U}, r(\cdot))$ which, as one would expect, given the development in Chapter II and the definition of symmetry isomorphism above, is a pair consisting of an open set $\mathcal{U}\subset\mathcal{B}$ and a smooth field $\underset{\sim}{r}(\cdot)$ of local reference configurations on $\mathcal{U}$ such that the symmetry groups $\mathcal{G}_{\underset{\sim}{r}(p)}(p)$ are independent of p, $\forall p\varepsilon\mathcal{U}$ (Note the difference: if $(\mathcal{U},\underset{\sim}{r}(\cdot))$ were an elastic reference chart we would require that the relative response functions $\underset{\sim}{S}_{\underset{\sim}{r}(p)}(F,\theta,p)$ be independent of p, $\forall p\varepsilon\mathcal{U}$; this would also imply that $(\mathcal{U},\underset{\sim}{r}(\cdot))$ is a symmetry chart but, as already noted, the converse is usually not true). If $(\mathcal{U},\underset{\sim}{r}(\cdot))$ is a <u>symmetry chart</u> then we set $\mathcal{G}_{\underset{\sim}{r}} = \mathcal{G}_{r(p)}(p)$, $\forall p\varepsilon\mathcal{U}$ and we call $\mathcal{G}_{\underset{\sim}{r}}$ the <u>symmetry group of</u> $\mathcal{U}$ <u>relative</u> to $\underset{\sim}{r}(\cdot)$. As was the case in Chapter II with smooth reference maps $\underset{\sim}{r}(\cdot)$ defined on reference neighborhoods $\mathcal{U}\subset\mathcal{B}$, the field of local configurations in a symmetry chart can not, in general, be extended to a smooth global field on $\mathcal{B}$.

Following the lead of the previous chapters, we call two symmetry charts $(\mathcal{U},\underset{\sim}{r}(\cdot))$ and $(\bar{\mathcal{U}},\bar{r}(\cdot))$ compatible if for all $p\varepsilon\mathcal{U}\cap\bar{\mathcal{U}}$

$$\underset{\sim}{S}_{\underset{\sim}{r}(p)}(\underset{\sim}{F},\theta,p) = \underset{\sim}{S}_{\bar{r}(p)}(\underset{\sim}{F},\theta,p), \quad \forall \underset{\sim}{F},\theta \qquad \text{(V-18)}$$

Clearly (V-18) implies that $\underset{\sim}{r}(p)^{-1} \circ \bar{r}(p) \; \varepsilon \; g(p)$, $\forall p\varepsilon\mathcal{U}\cap\bar{\mathcal{U}}$ and also that $\mathcal{G}_{\underset{\sim}{r}} \equiv \mathcal{G}_{\bar{\underset{\sim}{r}}}$; as should be expected by now we define a <u>symmetry</u> atlas to be a maximal collection $\underset{\sim}{\Lambda} = \{(\mathcal{U}_\alpha,\underset{\sim}{r}^a(\cdot)),\ \alpha\varepsilon I\}$ (17) of mutually compatible symmetry

(17) we have repeatedly used the same symbol I in specifying the index set in each of the atlases used in this treatise; in general, these index sets will be different of course.

charts such that $\bigcup_I \mathcal{U}_\alpha = \mathcal{B}$. Note that the <u>coordinate trans-formation</u> $\underset{\sim}{G}_{\alpha\beta} \equiv \underset{\sim}{r}^\alpha \circ \underset{\sim}{r}^{\beta-1}$, from $\underset{\sim}{r}^\alpha$ to $\underset{\sim}{r}^\beta$, is a smooth field on $\mathcal{U}_\alpha \cap \mathcal{U}_\beta$, for all $\alpha,\beta\varepsilon I$, which takes its values in $\underset{\sim}{G}_\Lambda \equiv \underset{\sim}{G}_{r}\alpha$, $\alpha\varepsilon I$; we call $\underset{\sim}{G}_\Lambda$ the symmetry group of $\mathcal{B}$ relative to the atlas $\underset{\sim}{\Lambda}$. Obviously, we can also define a relative response function $\underset{\sim}{S}_{\underset{\sim}{\Lambda}}(F,\theta,\cdot)$ on $\mathcal{B}$ with

$$\underset{\sim}{S}_{\underset{\sim}{\Lambda}}(\underset{\sim}{F},\theta,p) = \underset{\sim}{S}_{\underset{\sim}{r}^\alpha}(\underset{\sim}{F},\theta,p), \quad \forall p\varepsilon\mathcal{U}_\alpha, \quad \alpha\varepsilon I \tag{V-19}$$

however, this smooth distribution of response functions, $\underset{\sim}{S}_{\underset{\sim}{\Lambda}}$, now depends, in general, on $p\varepsilon\mathcal{B}$; this again points up the relative weakness of the condition that the tangent spaces of different points of $\mathcal{B}$ be connected via symmetry isomorphisms instead of material isomorphisms. In fact, a necessary and sufficient condition for a symmetry atlas $\underset{\sim}{\Lambda}$ to be a material atlas is that the distribution of relative response functions $\underset{\sim}{S}_{\underset{\sim}{\Lambda}}$ be a fixed function which is independent of the material points $p\varepsilon\mathcal{B}$. From the definitions stated above it is immediate that $G_{\underset{\sim}{\Lambda}}$ is the symmetry group for the distribution $S_{\underset{\sim}{\Lambda}}$ of response functions relative to $\underset{\sim}{\Lambda}$ and that the coordinate transformations $\underset{\sim}{G}_{\alpha\beta}$ satisfy the conditions delineated on page II-16.

<u>Exercise</u> 35 Show that for each $K\varepsilon GL(3)$, $\underset{\sim}{K}\circ\underset{\sim}{\Lambda} = \{(\mathcal{U}_\alpha, \underset{\sim}{K}\circ\underset{\sim}{r}^\alpha(\cdot)), \alpha\varepsilon I\}$ is a symmetry atlas for $\mathcal{B}$ if $\underset{\sim}{\Lambda}$ is and verify the transformation rules

$$\underset{\sim}{S}_{\underset{\sim}{K}\underset{\sim}{\Lambda}}(\underset{\sim}{F},\theta,p) = \underset{\sim}{S}_{\Lambda}(\underset{\sim}{F}\underset{\sim}{K},\theta,p), \quad \forall \underset{\sim}{F},\theta,p$$

$$G_{\underset{\sim}{k}\underset{\sim}{\Lambda}} = \underset{\sim}{K}G_{\underset{\sim}{\Lambda}}\underset{\sim}{K}^{-1}$$

so that $G_{\underset{\sim}{\Lambda}}$ is characterized by the condition $\underset{\sim}{K}\varepsilon G_{\underset{\sim}{\Lambda}} \Longleftrightarrow \underset{\sim}{K}\circ\underset{\sim}{\Lambda} = \underset{\sim}{\Lambda}$ and $\underset{\sim}{\Lambda}$ is characterized by the distribution of the relative response functions $\underset{\sim}{S}_{\underset{\sim}{\Lambda}}$, i.e., $\underset{\sim}{\Lambda} = \underset{\sim}{\bar{\Lambda}} \Longleftrightarrow \underset{\sim}{S}_{\underset{\sim}{\Lambda}} = S_{\bar{\Lambda}}$.

Now, as has been pointed out by Wang in [36], apart from the fact that the distribution of relative response functions $\underset{\sim}{S}_{\underset{\sim}{\Lambda}}$ depends on the material points $p\varepsilon\mathcal{B}$, a symmetry atlas $\underset{\sim}{\Lambda}$ is formally the same as the material atlases $\underset{\sim}{\Phi}$ which were treated in Chapter II; the geometric structure induced on a thermoelastic body by a symmetry atlas (i.e., $\mathcal{B}$ is a smooth thermoelastic body with <u>uniform symmetry</u>) is, therefore, the same as that which is induced on a smooth materially uniform elastic body by a material atlas. In particular we make

<u>Definition</u> V-2. A <u>symmetry connection</u> on $\mathcal{B}$ is an affine connection whose induced parallel transports are symmetry isomorphisms.

<u>Exercise</u> 36 Let $\underset{\sim}{\Lambda}$ be a symmetry atlas for $\mathcal{B}$ and let $(\mathcal{U},\underset{\sim}{r}(\cdot)) \varepsilon \underset{\sim}{\Lambda}$ be a symmetry chart. If $\kappa: \mathcal{B} \to R^3$ induces coordinates (x^i) on $\mathcal{B}$ and $\underset{\sim}{F} = \kappa_*\circ\underset{\sim}{r}^{-1}$ is the deformation gradient from $\underset{\sim}{r}$ to κ_*, prove that the functions Γ^i_{jk} are the connection symbols of a symmetry connection on $\mathcal{B}$ iff the fields of matrices

$$[F^{-1i}_{\ \ j}(\frac{\partial F^j_k}{\partial x^m} + \Gamma^j_{lm}F^l_k)], \quad m = 1,2,3$$

belong to $\underset{\sim}{g}_\Lambda$, the Lie algebra of the symmetry group $\underset{\sim}{G}_\Lambda$.
hint: follow the same argument as that used in proving Theorem II-3.

4. Thermodynamics and Anelastic Response

Let $\mathcal{B}$ be a material body and p a point of $\mathcal{B}$. Recall that, in the previous chapter, we have defined p to be an anelastic point if it is a quasi-elastic point such that the instanteous (anelastic) response function $\underset{\sim}{E}_p^t$ is given by transforming a fixed elastic response function $\underset{\sim}{E}_p$ according to the rule (IV-3) where $t_o \varepsilon R$, $\underset{\sim}{r}(p)$ is any element of $\mathcal{D}_p$, and $\underset{\sim}{\alpha}(p,t)$ is an isochoric automorphism of $\mathcal{B}_p$ called an anelastic transformation function; in any rigid or rest process of p we assumed that $\underset{\sim}{\alpha}(p,t) = id_{\mathcal{B}_p}$ and that $\underset{\sim}{\alpha}(p,t) \to id_{\mathcal{B}_p}$, as $t \to -\infty$, in any process of p. We also assumed in Chapter IV that the evolution of an anelastic transformation function $\underset{\sim}{\alpha}(p,t)$ is governed by a flow rule of the form $\dot{\underset{\sim}{\alpha}}(p,t) = \underset{\sim}{\psi}(\underset{\sim}{\alpha}(p,t), t)$ and we examined certain questions related to the uniqueness of such transformation functions.

In order to generalize the mechanical model of anelasticity, which was presented in Chapter IV, to a thermomechanical one we begin with the list, i.e. (V-1) - (V-4), of constitutive relations for a thermoelastic point p and regard the constitutive functions $\underset{\sim}{E}_p$, e_p, $\underset{\sim}{h}_p$, and $\underset{\sim}{\ell}_p$ as the elastic response functions which hold for a thermoanelastic point p in any rest process with constant $(\underset{\sim}{\nu}(p), \theta(p), \underset{\sim}{g}(p))$.

If the process is not a rest process then we assume that the instantaneous response functions $(\underset{\sim}{E}_p^t, e_p^t, h_p^t, \underset{\sim}{\ell}_p^t)$ are related to $(\underset{\sim}{E}_p, e_p, h_p, \underset{\sim}{\ell}_p)$ via

$$(\underset{\sim}{E}_p^t, e_p^t, h_p^t, \underset{\sim}{\ell}_p^t)(\underset{\sim}{\nu}(p),\ \theta(p),\ \underset{\sim}{g}(p)) \tag{V-20}$$

$$= (\underset{\sim}{E}_p, e_p, h_p, \underset{\sim}{\ell}_p)(\underset{\sim}{\nu}(p) \circ \underset{\sim}{\alpha}(p,t),\ \theta(p),\ \underset{\sim}{g}(p))$$

where $\underset{\sim}{\alpha}(p,t)$, the anelastic transformation function, is governed by the flow rule associated with the thermodynamic process to which $\mathcal{B}$ has been subjected.

We now want to determine the thermodynamical restrictions on the constitutive relations (V-20) which follow from the reduced entropy inequality (V-5). In order to proceed we again choose a fixed local reference configuration $\underset{\sim}{r}(p)$ in $\mathcal{D}_p$ and rewrite the system (V-20) in the form

$$(\underset{\sim}{S}^t_{\underset{\sim}{r}(p)},\ e^t_{\underset{\sim}{r}(p)},\ h^t_{\underset{\sim}{r}(p)},\ \underset{\sim}{\ell}^t_{\underset{\sim}{r}(p)})(\underset{\sim}{F},\theta,\underset{\sim}{g},p) \tag{V-20'}$$

$$= (\underset{\sim}{S}_{\underset{\sim}{r}(p)},\ e_{\underset{\sim}{r}(p)},\ h_{\underset{\sim}{r}(p)},\ \underset{\sim}{\ell}_{\underset{\sim}{r}(p)})(\underset{\sim}{F}\underset{\sim}{A}_{\underset{\sim}{r}(p)}(t),\theta,\underset{\sim}{g},p)$$

where $\underset{\sim}{F} = \underset{\sim}{\nu}(p) \circ \underset{\sim}{r}(p)^{-1}$ is the deformation gradient relative to $\underset{\sim}{r}(p)$ and $A_{\underset{\sim}{r}(p)}(t) = \underset{\sim}{r}(p) \circ \underset{\sim}{\alpha}(p,t) \circ \underset{\sim}{r}(p)^{-1}$ is the anelastic transformation function relative to $r(p)$; as $\underset{\sim}{r}(p)$, the preferred local reference configuration of p which we have chosen, will be held fixed in the analysis to follow, we will now suppress the $\underset{\sim}{r}(p)$ subscripts which appear in the response functions and the relative anelastic transformation function. The free energy function $\psi \equiv \varepsilon - \eta\theta$ then has the form

$$\psi = \mathcal{F}(F, \theta, g, p) \tag{V-21}$$

in any rest process and

$$\psi = \mathcal{F}^t(\underset{\sim}{F}, \theta, \underset{\sim}{g}, p) \equiv \mathcal{F}(\underset{\sim}{F}\underset{\sim}{A}(t), \theta, \underset{\sim}{g}, p) \tag{V-22}$$

at any $t \in \mathcal{R}$ in a general thermodynamic process. From (V-22) we easily get

$$\frac{\partial \mathcal{F}^t}{\partial F^i_j} = \frac{\partial \mathcal{F}}{\partial F^i_k} A^j_k(t), \quad \frac{\partial \mathcal{F}^t}{\partial \theta} = \frac{\partial \mathcal{F}}{\partial \theta}, \quad \frac{\partial \mathcal{F}^t}{\partial g_i} = \frac{\partial \mathcal{F}}{\partial g_i} \tag{V-23}$$

as the relations which exist among the various gradients of $\mathcal{F}$ and $\mathcal{F}^t$. A direct computation employing (V-22) and (V-23) now yields

$$\dot{\psi} = \frac{\partial \mathcal{F}}{\partial F^a_b} (\dot{F}^a_c A^c_b + F^a_c \dot{A}^c_b) + \frac{\partial \mathcal{F}}{\partial \theta} \dot{\theta} \tag{V-24}$$

$$+ \frac{\partial \mathcal{F}}{\partial g_a} \dot{g}_a$$

$$= \frac{\partial \mathcal{F}^t}{\partial F^a_b} \dot{F}^a_b + \frac{\partial \mathcal{F}^t}{\partial F^a_b} F^a_c \dot{A}^c_d A^{-1d}_{\ \ \ c}$$

$$+ \frac{\partial \mathcal{F}^t}{\partial \theta} \dot{\theta} + \frac{\partial \mathcal{F}^t}{\partial g_a} \dot{g}_a$$

By substituting (V-24) into (V-5) we easily obtain the following results

I. both F and F^t are independent of $\underset{\sim}{g}$, i.e.,

$\psi = F^t(\underset{\sim}{F},\theta,p) = F(\underset{\sim}{F}\underset{\sim}{A}(t),\theta,p)$

II. both h and h^t are independent of $\underset{\sim}{g}$ and are related to F and F^t via

$$h^t(\underset{\sim}{F},\theta,p) = h(\underset{\sim}{F}\underset{\sim}{A}(t),\theta,p)$$

$$= -\frac{\partial F}{\partial\theta}(\underset{\sim}{F}\underset{\sim}{A}(t),\ \theta,p)$$

$$= -\frac{\partial F^t}{\partial\theta}\ (\underset{\sim}{F},\theta,p)$$

III. $\underset{\sim}{S}$ and $\underset{\sim}{S}^t$ are both independent of $\underset{\sim}{g}$ and are related to F and F^t via

$$\underset{\sim}{S}{}^{ta}_{b}(\underset{\sim}{F},\theta,p) = S^a_b(\underset{\sim}{F}\underset{\sim}{A}(t),\theta,p)$$

$$= \rho F^a_c A^c_d(t)\ \frac{\partial F(FA(t),\theta,p)}{\partial F^b_d}$$

$$= \rho F^a_c\ \frac{\partial F^t(F,\theta,p)}{\partial F^b_c}$$

IV. $\underset{\sim}{\ell}$ and $\underset{\sim}{\ell}^t$ satisfy the inequality

$$\ell^j(\underset{\sim}{F}\underset{\sim}{A}(t),\theta,\underset{\sim}{g},p)g_j = \ell^{tj}(\underset{\sim}{F},\theta,\underset{\sim}{g},p)g_j \le 0$$

and

V. $\mathring{A}(t)$ satisfies the inequality

$$F^{-1a}_{\ \ b}\ S^{tb}_c(\underset{\sim}{F},\theta,\underset{\sim}{g})F^c_d\mathring{A}^d_e(t)A^{-1e}_{\ \ a} \le 0$$

Exercise 37. Preform the substitution of (V-24) into (V-5) and verify the results I through V above.

Remarks If we compare the results given above, i.e., I-V, with (V-6) through (V-9) of §2, we easily see that the elastic response functions $\underset{\sim}{S}$, e, h, and $\underset{\sim}{\ell}$ of a thermo-anelastic point p satisfy exactly the same thermodynamic restrictions as those satisfied by the response functions of a thermoelastic point. The complete constitutive theory for a thermoanelastic point p now consists of the constitutive relations (V-20)(or (V-1) through (V-4)), the thermodynamic restrictions on the anelastic response functions which are given by I-IV above, the flow rule, and the restriction on the flow rate which is contained in condition V above.

5. Symmetry Groups and Symmetry Isomorphisms in Thermo-Anelasticity.

Because the elastic response functions of a thermo-anelastic point obey the same restrictions as those of a thermoelastic point we define the elastic symmetry groups by

$$\mathcal{g}(p) = \{\underset{\sim}{A} \varepsilon SL(\mathcal{B}_p) \mid (\underset{\sim}{E}_p, \underset{\sim}{\ell}_p)(\underset{\sim}{\nu}(p) \circ \underset{\sim}{A}, \theta(p), \underset{\sim}{g}(p)) = (\underset{\sim}{E}_p, \underset{\sim}{\ell}_p)(\underset{\sim}{\nu}(p), \theta(p), \underset{\sim}{g}(p)), \forall \underset{\sim}{\nu}, \theta, \underset{\sim}{g}\}$$

and

$$\mathcal{g}^*(p) = \{\underset{\sim}{A} \varepsilon SL(\mathcal{B}_p) \mid (e, h, \psi)(\underset{\sim}{\nu}(p) \circ \underset{\sim}{A}, \theta(p)) = (e, h, \psi)(\underset{\sim}{\nu}(p), \theta(p)), \forall \underset{\sim}{\nu}, \theta\}$$

assuming, as we have already indicated, that the symmetry group of $\underset{\sim}{\ell}_p$, for each $p\epsilon\mathcal{B}$, coincides with that of $\underset{\sim}{E}_p$. The elastic symmetry groups $g(p)$ and $g^*(p)$ give rise, of course, to the anelastic symmetry groups $g^t(p)$ and $g^{*t}(p)$ via

$$g^t(p) = \underset{\sim}{\alpha}(p,t) \circ g(p) \circ \underset{\sim}{\alpha}(p,t)^{-1} \tag{V-25}$$

and

$$g^{*t}(p) = \underset{\sim}{\alpha}(p,t) \circ g^*(p) \circ \alpha(p,t)^{-1} \tag{V-26}$$

<u>Exercise</u> 38 Use (V-25) and (V-26) to prove that

$$\underset{\sim}{A}\epsilon g^t(p) \Longleftrightarrow \underset{\sim}{A}\epsilon SL(\mathcal{B}_p) \text{ and}$$

$$\{(\underset{\sim}{E}{}^t_p,\underset{\sim}{\ell}{}^t_p)(\underset{\sim}{\nu}(p) \circ \underset{\sim}{A},\ \theta(p),\ \underset{\sim}{g}(p))$$

$$= (\underset{\sim}{E}{}^t_p,\underset{\sim}{\ell}{}^t_p)(\nu(p),\ \theta(p),\ \underset{\sim}{g}(p)),$$

$$\forall \underset{\sim}{\nu},\theta,\underset{\sim}{g}\}$$

and that

$$\underset{\sim}{A}\epsilon g^{*t}(p) \Longleftrightarrow \underset{\sim}{A}\epsilon SL(\mathcal{B}_p) \text{ and}$$

$$\{(e^t_p,h^t_p,\mathfrak{f}^t_p)(\underset{\sim}{\nu}(p) \circ A,\ \theta(p))$$

$$= (e^t_p,h^t_p,\mathfrak{f}^t_p)(\underset{\sim}{\nu}(p),\ \theta(p)),\quad \forall \underset{\sim}{\nu},\theta\}$$

where $\mathfrak{f}_p(\underset{\sim}{\nu},\theta) = F(\underset{\sim}{\nu}(p) \circ \underset{\sim}{r}(p)^{-1},\ \theta(p),\ p)$ and

$\mathfrak{f}^t_p(\underset{\sim}{\nu}(p),\ \theta(p)) \equiv \mathfrak{f}_p(\underset{\sim}{\nu}(p) \circ \underset{\sim}{\alpha}(p,t),\ \theta(p)).$

Relative to the preferred local reference configuration $\underset{\sim}{r}(p) \in \mathcal{D}_p$ which we have singled out, the groups $g(p)$, $g^*(p)$, $g^t(p)$, and $g^{*t}(p)$ are represented by the symmetry groups $G_{\underset{\sim}{r}(p)}(p)$, $G^*_{\underset{\sim}{r}(p)}(p)$, $G^t_{\underset{\sim}{r}(p)}(p)$, and $G^{*t}_{\underset{\sim}{r}(p)}(p)$; the first two of these symmetry groups are given by (V-14) and (V-15), respectively, while

$$G^t_{\underset{\sim}{r}(p)}(p) = \underset{\sim}{r}(p) \circ g^t(p) \circ \underset{\sim}{r}(p)^{-1}$$

$$G^{*t}_{\underset{\sim}{r}(p)}(p) = \underset{\sim}{r}(p) \circ g^{*t}(p) \circ \underset{\sim}{r}(p)^{-1}$$

so that

$$\underset{\sim}{K} \in G^t_{\underset{\sim}{r}(p)}(p) \iff \underset{\sim}{K} \in SL(3) \text{ and}$$

$$\{(\underset{\sim}{S}^t, \underset{\sim}{\ell}^t)(\underset{\sim}{F}\underset{\sim}{K}, \theta, g) = (\underset{\sim}{S}^t, \underset{\sim}{\ell}^t)(\underset{\sim}{F}, \underset{\sim}{\theta}, \underset{\sim}{g}), \ \forall (\underset{\sim}{F}, \theta, \underset{\sim}{g})$$

with a similar result for $G^{*t}_{r(p)}(p)$.

Having considered the constitutive relations and symmetry groups of an arbitrary thermoanelastic point $p \in \mathcal{B}$ we will now assume that all the body points of $\mathcal{B}$ are thermoanelastic and that $\mathcal{B}$ satisfies the following global constitutive assumptions which were delineated in [20]:

A. the elastic response functions of the body points are distrubuted on $\mathcal{B}$ in exactly the same way as those of a thermoelastic body with uniform symmetry, i.e., in all rest processes the global structure of $\mathcal{B}$ is precisely the same as that which was formulated in §3 of this chapter.

B. in a general procsss of $\mathcal{B}$ the elastic global structure evolves smoothly into the anelastic global structure, through the anelastic transformation functions $\underset{\sim}{\alpha}$, in exactly the same way as in the (mechanical) anelasticity theory of Chapter IV.

Within the context of the present situation, hypotheses A and B above imply all of the following: First of all, there exists an <u>elastic</u> <u>symmetry</u> atlas $\underset{\sim}{\Lambda}$ for B which is a collection $\underset{\sim}{\Lambda} = \{(\mathcal{U}_\alpha, \underset{\sim}{r}^\alpha), \alpha\varepsilon I\}$ with $\mathcal{U}_\alpha$, an open set in $\mathcal{B}$, for each $\alpha\varepsilon I$, such that $\underset{I}{\cup}\mathcal{U}_\alpha = \mathcal{B}$, and $\underset{\sim}{r}^\alpha(\cdot)$ a smooth field of local reference configurations on $\mathcal{U}_\alpha$ which satisfies

(i) there exists a smooth distribution of relative response functions $(\underset{\sim}{S}_{\underset{\sim}{\Lambda}}, \underset{\sim}{\ell}_{\underset{\sim}{\Lambda}})$ on $\mathcal{B}$ such that

$$(\underset{\sim}{S}_{\underset{\sim}{\Lambda}}, \underset{\sim}{\ell}_{\underset{\sim}{\Lambda}})(\underset{\sim}{F}, \theta, \underset{\sim}{g}, p) = (\underset{\sim}{S}_{\underset{\sim}{r}^\alpha}, \underset{\sim}{\ell}_{\underset{\sim}{r}^\alpha})(F, \theta, \underset{\sim}{g}, p) \qquad \text{(V-27)}$$

$\forall p\varepsilon\mathcal{U}_\alpha$, $\forall\alpha\varepsilon I$.

(ii) the distribution $(\underset{\sim}{S}_{\underset{\sim}{\Lambda}}, \underset{\sim}{\ell}_{\underset{\sim}{\Lambda}})$ has a <u>uniform</u> <u>relative</u> <u>symmetry</u> <u>group</u> $G_{\underset{\sim}{\Lambda}}$, i.e.,

$$G_{\underset{\sim}{\Lambda}} = G_{\underset{\sim}{r}^\alpha(p)}(p), \quad \forall p\varepsilon\mathcal{U}_\alpha, \quad \forall\alpha\varepsilon I \qquad \text{(V-28)}$$

Secondly, the anelastic global structure of $\mathcal{B}^{\cdot}$ at any time t in a general process is obtained from the elastic global structure as follows: there exists a smooth field of global anelastic transformation functions $\underset{\sim}{\alpha}(\cdot,t)$ on $\mathcal{B}$ such that an <u>anelastic symmetry atlas</u> $\underset{\sim}{\Lambda}(t)$ may be defined by

$\underset{\sim}{\Lambda}(t) = \{(\mathcal{U}_\alpha, \underset{\sim}{r}^\alpha(\cdot,t)), \quad \alpha\varepsilon I\}$ where for each $\alpha\varepsilon I$, $\underset{\sim}{r}^\alpha(\cdot,t) = \underset{\sim}{r}^\alpha(\cdot) \circ \underset{\sim}{\alpha}(\cdot,t)^{-1}$ on $\mathcal{U}_\alpha$, $\forall t\varepsilon R$; this anelastic symmetry atlas satisfies

(i') there exists a smooth distribution of relative response functions $(\underset{\sim}{S}^t_{\underset{\sim}{\Lambda}(t)}, \underset{\sim}{\ell}^t_{\underset{\sim}{\Lambda}(t)})$ on $\mathcal{B}$ such that

$$(\underset{\sim}{S}^t_{\underset{\sim}{\Lambda}(t)}, \underset{\sim}{\ell}^t_{\underset{\sim}{\Lambda}(t)})(\underset{\sim}{F},\theta,\underset{\sim}{g},p) \tag{V-29}$$

$$= (\underset{\sim}{S}^t_{\underset{\sim}{r}^\alpha(p,t)}, \underset{\sim}{\ell}^t_{\underset{\sim}{r}^\alpha(p,t)})(\underset{\sim}{F},\theta,\underset{\sim}{g},p)$$

$\forall p\varepsilon\mathcal{U}_\alpha$, $\forall\alpha\varepsilon I$.

(ii') the symmetry group of $(\underset{\sim}{S}^t_{\underset{\sim}{\Lambda}(t)}, \underset{\sim}{\ell}^t_{\underset{\sim}{\Lambda}(t)})$ is uniform on $\mathcal{B}$, i.e.,

$$G^t_{\underset{\sim}{\Lambda}(t)} = G^t_{\underset{\sim}{r}^\alpha(t,p)}(p), \quad \forall p\varepsilon\mathcal{U}_\alpha, \ \alpha\varepsilon I \tag{V-30}$$

<u>Remarks</u> Note that, just as in Chapter IV, the relative anelastic response functions coincide with the relative elastic response functions, i.e., $(\underset{\sim}{S}^t_{\underset{\sim}{\Lambda}(t)}, \underset{\sim}{\ell}^t_{\underset{\sim}{\Lambda}(t)}) \equiv (\underset{\sim}{S}_{\underset{\sim}{\Lambda}}, \underset{\sim}{\ell}_{\underset{\sim}{\Lambda}})$ and, moreover, the relative anelastic symmetry group $\tilde{G}^t_{\underset{\sim}{\Lambda}(t)}$ coincides with $G_{\underset{\sim}{\Lambda}}$; these facts are, of course, a simple consequence of the fact that the elastic and anelastic symmetry charts $(\mathcal{U}_\alpha, \underset{\sim}{r}^\alpha(\cdot))$ and $(\mathcal{U}_\alpha, \underset{\sim}{r}^\alpha(\cdot,t))$, respectively,

are connected via $\underset{\sim}{r}^{\alpha}(\cdot,t) = \underset{\sim}{r}^{\alpha}(\cdot) \circ \underset{\sim}{\alpha}(\cdot,t)^{-1}$, $\forall t\varepsilon R$, $\forall p\varepsilon U_{\alpha}$. Note also that, as in §3 of this chapter, the relative response functions $\underset{\sim}{S}_{\underset{\sim}{\Lambda}}$ and $S^{t}_{\underset{\sim}{\Lambda}(t)}$ depend explicitly on the points $p\varepsilon B$ since we are working with the concept of symmetry isomorphism rather than with that of material isomorphism.

6. Structural Connections on Thermo-anelastic Bodies

Just as in §3 we now define an elastic symmetry connection on a thermoanelastic body B to be an affine connection on the tangent bundle of B whose induced parallel transports of the tangent spaces along any smooth curve $\lambda \subset B$ are symmetry isomorphisms. If we denote these parallel transports by $\rho(\tau)$ and let $(U_{\alpha},\underset{\sim}{r}^{\alpha}(\cdot))$ be a symmetry chart in $\underset{\sim}{\Lambda}$, an elastic symmetry atlas for B, with $\lambda \subset U_{\alpha}$, then we know that $\rho(\tau)$ is a symmetry isomorphism iff the induced mappings $\rho_{t,\alpha} \equiv \underset{\sim}{r}_{\alpha}(\lambda(t)) \circ \rho \circ \underset{\sim}{r}_{\alpha}^{-1}(\lambda(0))$ are elements of $G_{\underset{\sim}{\Lambda}}$ which pass through the identity element at t=0; this latter condition implies that the vector $\frac{d}{dt}\rho_{t,\alpha}\big|_{t=0} \varepsilon \underset{\approx}{g}_{\underset{\sim}{\Lambda}}$. We have already demonstrated that if we represent the charts $(U_{\alpha},\underset{\sim}{r}^{\alpha}(\cdot))$ in $\underset{\sim}{\Lambda}$ by the deformation gradient $\underset{\sim}{F} \equiv \kappa_{*}\circ \underset{\sim}{r}^{\alpha-1}$, where $\kappa\colon B \to R^{3}$ induces a global coordinate system (X^{A}) on B, then a necessary and sufficient condition for the functions $\Gamma^{A}_{BC}(X^{D})$ to represent the connection symbols of a symmetry connection H on B is that

$$[F^{-1A}_{\;\;C}(\frac{\partial F^{C}_{B}}{\partial X^{D}} + \Gamma^{C}_{ED}F^{E}_{B})] \;\varepsilon\; \underset{\approx}{g}_{\underset{\sim}{\Lambda}}\;,\; D = 1,2,3.$$

Recall now that in Chapter IV we were able to generate an anelastic material connection H^t from an elastic material connection H by relating the parallel transports $\rho(\tau)$ and $\rho^t(\tau)$ of H and H^t, respectively, along any path $\lambda \subset \mathcal{B}$ from $\lambda(0)$ to $\lambda(\tau)$, by

$$\rho^t(\tau) = \underset{\sim}{\alpha}(\lambda(\tau),t\ ) \circ \rho(\tau) \circ \underset{\sim}{\alpha}(\lambda(0),t)^{-1} \tag{V-31}$$

where $\underset{\sim}{\alpha}(\cdot,t)$ is some field of anelastic transformation functions (in a general process of $\mathcal{B}$) which maps the elastic symmetry atlas $\underset{\sim}{\Lambda}$ of the elastic structure to the anelastic symmetry atlas $\underset{\sim}{\Lambda}(t)$ of the anelastic structure. In the same way, we may use the relation (V-31) to generate an <u>anelastic symmetry connection</u> from an elastic symmetry connection. In other words, if H is a structural connection relative to $\underset{\sim}{\Lambda}$ which induces parallel transports $\rho(\tau)$, then the connection H^t whose induced parallel transports $\rho^t(\tau)$ are given via (V-31) is a structural connection relative to $\underset{\sim}{\Lambda}(t)$, where corresponding charts $(\mathcal{U}_\alpha, r^\alpha(\cdot))$ and $(\mathcal{U}_\alpha, \underset{\sim}{r}^\alpha(\cdot,t))$ in $\underset{\sim}{\Lambda}$ and $\underset{\sim}{\Lambda}(t)$, respectively, are related by (IV-39) with $\underset{\sim}{\alpha}(\cdot,t)$ the field of anelastic transformation functions which appears in (V-31). We may, therefore, state the following

<u>Theorem</u> V-1 Let $\underset{\sim}{\Lambda}$ and $\underset{\sim}{\Lambda}(t)$ be, respectively, elastic and anelastic symmetry atlases for $\mathcal{B}$ which are related via (IV-39). Then the functions $\Gamma^A_{BC}\ (X^D,t)$ are the connection

symbols of an anelastic symmetry connection on $\mathcal{B}$, relative to $\underset{\sim}{\Lambda}(t)$, iff the matrices

$$[F^{-1A}_{\ C}(X^F,t)(\frac{\partial F^C_B(X^F,t)}{\partial X^D} + \Gamma^C_{ED}(X^F,t)F^E_B(X^F,t))],$$

$D = 1,2,3$, are contained in $\underset{\sim}{g}_{\Lambda(t)} \equiv \underset{\sim}{g}_{\Lambda}$, where the functions $F^A_B(X^C,t)$ are the components of the deformation gradient $F_\alpha \equiv \kappa_* \circ \underset{\sim}{r}^\alpha(\cdot,t)^{-1}$ from $\underset{\sim}{r}^\alpha$ to κ_*. Moreover, the functions $\Gamma^A_{BC}(\cdot,t)$ and the connection symbols $\Gamma^A_{BC}(\cdot)$ of the associated elastic symmetry connection relative to $\underset{\sim}{\Lambda}$ are related by (IV-50).

7. Field Equations for Thermoelastic and Thermoanelastic Bodies.

The procedure which we will follow in order to derive global field equations for thermoelastic bodies with uniform symmetry is not very different from that which was employed in Chapter II, where we derived global equations of motion for materially uniform elastic bodies. Of course, as we are dealing here with a thermodynamical theory, we must take into account not only the equations of motion which are to be satisfied by points of $\mathcal{B}$ but also the equation of heat transfer; once the appropriate global field equations for thermoelastic bodies have been found the approach followed in §8 of Chapter IV will produce the corresponding global field equations which apply to thermoanelastic bodies exhibiting uniform symmetry.

We begin by considering the equations of motion for a thermoelastic body $\mathcal{B}$ and assume that we have singled out a fixed elastic symmetry atlas $\underset{\sim}{\Lambda} = \{(\mathcal{U}_\alpha, \underset{\sim}{r}^\alpha(\cdot)),\ \alpha\varepsilon I\}$; in what follows we will suppress the subscript $\underset{\sim}{\Lambda}$ on the relative response functions but we will retain the $\underset{\sim}{\Lambda}$ subscript on $\underset{\sim}{G}_\Lambda$ and $\underset{\approx}{g}_\Lambda$. Now let $\phi(t)\colon \mathcal{B} \to R^3$ be an arbitrary motion of $\mathcal{B}$ which induces time-dependent coordinates $(x^i(t))$ on $\mathcal{B}$. If we denote by $\phi_t(\mathcal{B})$ the configuration of $\mathcal{B}$ at time t in the motion ϕ, then the stress tensor in $\phi_t(\mathcal{B})$ is determined, of course, as follows: for $p\varepsilon\mathcal{B}$ we choose $(\mathcal{U}_\alpha, \underset{\sim}{r}^\alpha(\cdot)) \ \varepsilon\ \underset{\sim}{\Lambda}$ such that $p\varepsilon\mathcal{U}_\alpha$ and we set $\underset{\sim}{F}_p = \phi_{*p}(t) \circ \underset{\sim}{r}^\alpha(p)^{-1}$. Then, relative to the standard basis of R^3, the components of the Cauchy stress tensor at p, at time t, are given by

$$T^{ij}(\phi_t(p)) = S^{ij}(\underset{\sim}{F}_p,\ \theta(p), p) \tag{V-32}$$

which is, of course, a local formula valid only for points $p\varepsilon\mathcal{U}_\alpha$ (note that $[\phi_t(p)]^i = x^i(p)$).

<u>Exercise</u> 39 If, as usual, we set

$$H^{ij}_{kl}(\underset{\sim}{F},\theta,p) \equiv \partial S^{ij}/\partial F^{kl}$$

for all $\underset{\sim}{F}\varepsilon GL(3)$, where the F^{kl} are the components of $\underset{\sim}{F}$ relative to (x^i) then

$$H^{ij}_{kl}(\underset{\sim}{F},\theta,p)F^{km}K^l_m = 0, \quad \forall \underset{\sim}{K}\varepsilon \underset{\approx}{g}_{\underset{\sim}{\Lambda}}$$

and

$$H^{ij}_{kl}(\underset{\sim}{F}\underset{\sim}{G},\theta,p)G^{l}_{r} = H^{ij}_{kr}(\underset{\sim}{F},\theta,p), \quad \forall \underset{\sim}{G}\varepsilon \underset{\sim}{G_{\Lambda}}$$

are both consequences of the fact that

$$\underset{\sim}{S}(\underset{\sim}{F}\underset{\sim}{G},\theta,p) = \underset{\sim}{S}(\underset{\sim}{F},\theta,p), \quad \forall(\underset{\sim}{F},\theta,p)$$

if $\underset{\sim}{G}\varepsilon \underset{\sim}{G_{\Lambda}}$. Prove these statements by following the same procedure as that which was employed in Chapter II, §9.

Now, from the substitution of (V-32) into Cauchy's equation of motion (II-24) we get

$$H^{ij}_{kl}\frac{\partial F^{kl}}{\partial x_{j}} + H^{ij}_{\theta}g_{j} + H^{i} + \rho b^{i} = \rho \ddot{x}^{i} \tag{V-33}$$

where $H^{ij}_{\theta} \equiv \partial S^{ij}/\partial\theta$ and $H^{i} = \partial S^{ij}/\partial x^{j}$. The arguments of H^{ij}_{kl}, H^{ij}_{θ}, and H^{i} are $\underset{\sim}{F}$, $\theta(p)$, and x^{m} , of course, and like H^{ij}_{kl} the fields H^{ij}_{θ} and H^{i} satisfy the symmetry conditions

$$H^{ij}_{\theta}(\underset{\sim}{F}\underset{\sim}{G},\theta,x^{m}) = H^{ij}_{\theta}(\underset{\sim}{F},\theta,x^{m}), \quad \forall \underset{\sim}{G}\varepsilon \underset{\sim}{G_{\Lambda}}$$

and

$$H^{i}(\underset{\sim}{F}\underset{\sim}{G},\theta,x^{m}) = H^{i}(\underset{\sim}{F},\theta,x^{m}), \quad \forall \underset{\sim}{G}\varepsilon \underset{\sim}{G_{\Lambda}}$$

for all $(\underset{\sim}{F},\theta,p)$.

Exercise 40 Prove that the local fields H^{ij}_{θ} and H^{i} are independent of the choice of the elastic symmetry chart $(\mathcal{U}_{\alpha},\underset{\sim}{r}^{\alpha}(\cdot)) \;\varepsilon\; \underset{\sim}{\Lambda}$.

As a direct consequence of the above exercise, it follows that only the term $H^{ij}_{kl}\dfrac{\partial F^{kl}}{\partial x^j}$, in the equations of motion (V-33), is a local expression; however, if we choose an elastic symmetry connection H on $\mathcal{B}$ whose connection symbols relative to (x^i) are functions $\Gamma^i_{jk}(x^m)$ then, because the matrices

$$[F^{-1i}_{j}(\frac{\partial F^i_k}{\partial x^m} + \Gamma^j_{lm}F^l_k)],\ m = 1,2,3$$

must lie in $\underset{\sim}{g}_{\underset{\sim}{\Lambda}}$, it follows that

$$H^{ij}_{kl}(\frac{\partial F^{kl}}{\partial x^m} + \Gamma^k_{nm}F^{nl}) = 0 \tag{V-34}$$

We can, therefore, rewrite (V-33) in the form

$$- H^{ij}_{kl}F^{nl}\Gamma^k_{nj} + H^{ij}_{\theta}g_j + H^i + \rho b^i = \rho\ddot{x}^i \tag{V-35}$$

<u>Exercise</u> 41 Show that the field

$$\bar{H}^{ijn}_{k} \equiv H^{ij}_{kl}(\underset{\sim}{F},\theta,x^m)F^{nl}$$

is independent of the choice of the elastic symmetry chart $(\mathcal{U}_\alpha,\underset{\sim}{r}^\alpha(\cdot)) \in \underset{\sim}{\Lambda}$ and thus prove that the equation of motion (V-35) is a global equation valid for all points $p\in\mathcal{B}$.

Now, as in Chapter II, the equation of motion (V-35) is not a convenient one with which to work since $\Gamma^i_{jk} = \Gamma^i_{jk}(x^m)$ and x^m varies with time in a motion $\phi(t)$ of $\mathcal{B}$. Thus, we follow the lead of Chapter II and introduce a fixed reference configuration $\kappa: \mathcal{B} \to R^3$ which induces coordinates (X^A) on $\mathcal{B}$. The motion $\phi(t)$ is, once again, characterized by the deformation functions $x^i = x^i(X^A,t)$, while the connection symbols $\overset{\kappa}{\Gamma}{}^A_{BC}$ of H, taken relative to (X^A), and the components $\overset{\kappa}{F}{}^{Aj}$ of $\overset{\kappa}{\underset{\sim}{F}}_\alpha \equiv \kappa_* \circ \underset{\sim}{r}^{\alpha-1}$ are related, respectively, to the functions Γ^i_{jk} and F^{ij} by the equations preceding (II-29); substituting from these equations into (V-35) we get

$$\tilde{H}^{ijA}_k \left(\frac{\partial^2 x^k}{\partial X^A \partial X^B} - \overset{\kappa}{\Gamma}{}^C_{AB} \frac{\partial x^k}{\partial X^C}\right) \frac{\partial X^B}{\partial x^j} \tag{V-36}$$

$$+ H^{ij}_\theta g_j + H^i + \rho b^i = \rho \ddot{x}^i$$

where $\tilde{H}^{ijA}_k \equiv H^{ij}_{kl}(\underset{\sim}{F},\theta,X^D)\overset{\kappa}{F}{}^{Al}$.

<u>Remarks</u> It might be instructive at this point to indicate exactly how one would arrive at the appropriate field equations of motion in terms of the Piola-Kirchhoff stress tensor $\overset{\kappa}{\underset{\sim}{T}}$ (which is related to the Cauchy stress tensor $\underset{\sim}{T}_S$ via $\overset{\kappa}{T}{}^A_k = J T^l_k \frac{\partial X^A}{\partial x^l}$, $J = \det[x^i{}_{,A}]$). We begin by defining the Piola-Kirchoff response function $\underset{\sim}{P}$ relative to $\underset{\sim}{\Lambda}$ (actually, $\underset{\sim}{P}_{\underset{\sim}{\Lambda}}$) by

$$\underset{\sim}{P}(\underset{\sim}{F},\theta,p) \equiv (\det \underset{\sim}{F})\underset{\sim}{S}(\underset{\sim}{F},\theta,p)(\underset{\sim}{F}^{-1})^T \tag{V-37}$$

where the superscript T denotes "transpose". Solving (V-37) for $\underset{\sim}{S}$ we obtain

$$\underset{\sim}{S} = \frac{1}{(\det \underset{\sim}{F})} \underset{\sim}{P}(\underset{\sim}{F},\theta,p)\underset{\sim}{F}^T \tag{V-38}$$

whose component form, when differentiated with respect to $\underset{\sim}{F}$, yields (i.e., see [36])

$$H^{ij}_{kl} = \frac{1}{(\det \underset{\sim}{F})}\left[F^{j}_{m}Q^{im}_{kl} + (\delta^{j}_{k}\delta_{im} - F^{j}_{m}F^{-1}_{lk})P^{im}\right] \tag{V-39}$$

with $Q^{ij}_{kl} \equiv \partial P^{ij}/\partial F^{kl}$. Substituting (V-39) into (V-36) then yields

$$\tilde{Q}^{iA}_{kB}\left(\frac{\partial^2 x^k}{\partial X^A \partial X^B} - \overset{\kappa}{\Gamma}{}^{C}_{BA}\frac{\partial x^k}{\partial X^C}\right) - \overset{\kappa}{T}{}^{iA}\,\overset{\kappa}{\theta}{}^{B}_{AB} \tag{V-40}$$

$$+ JH^{ij}_{\theta}g_j + JH^i + \rho_\kappa b^i = \rho_\kappa \ddot{x}^i$$

where

$$\tilde{Q}^{iA}_{kB} = \frac{1}{(\det \overset{\kappa}{\underset{\sim}{F}})} Q^{ir}_{kl}(\underset{\sim}{F},\theta,X^D)\overset{\kappa}{F}{}^{A}_{r}\overset{\kappa}{F}{}^{Bl} \tag{V-41}$$

$$\overset{\kappa}{\theta}{}^{A}_{BC} = \overset{\kappa}{\Gamma}{}^{A}_{BC} - \overset{\kappa}{\Gamma}{}^{A}_{CB} \tag{V-42}$$

and

$$\overset{\kappa}{T}{}^{iA} = \frac{1}{(\det \overset{\kappa}{\underset{\sim}{F}})} P^{ik}(\underset{\sim}{F},\theta,X^D)\overset{\kappa}{F}{}^{A}_{k} \tag{V-43}$$

$$= JS^{ij}(\underset{\sim}{F},\theta,x^m)\,\frac{\partial X^A}{\partial x^j}$$

Note also that $JH^{ij}_{\theta}g_j = \frac{\partial \overset{\kappa}{T}{}^{iA}}{\partial \theta}\,\overset{\kappa}{g}_A$ and

$$JH^i = \frac{\partial \overset{\kappa}{T}{}^{iA}}{\partial X^A} = \frac{\partial \overset{\kappa}{T}{}^{iA}}{\partial x^m}\,x^m_{,A} \text{ where } \overset{\kappa}{g}_A = g_i x^i_{,A}$$

Now, let us turn our attention to the thermodynamical aspects of the global field equations for a thermoelastic body with uniform symmetry, i.e., to the derivation of a global form for the usual <u>equation of heat transfer</u>,

$$\operatorname{div} \underset{\sim}{q} + \rho r = \rho\theta\dot{\eta} \qquad \text{(V-44)}$$

where r respresents the energy supply.

<u>Remarks</u> the equation of heat transfer (V-44) arises in the following way: first of all we have the principle of balance of energy whose local field equation has the form

$$\rho\dot{\varepsilon} = \operatorname{tr}(\underset{\sim}{T}_S \operatorname{grad} \underset{\sim}{v}) + \operatorname{div} \underset{\sim}{q} + \rho r \qquad \text{(V-45)}$$

where $\underset{\sim}{v} = \dot{\underset{\sim}{x}}$ is the velocity field. Since $\underset{\sim}{T}_S$ and ε are determined by the free-energy function ψ via (V-7) and (V-8) we have

$$\rho\dot{\psi} \equiv \operatorname{tr}(\underset{\sim}{T}_S \operatorname{grad} \underset{\sim}{v}) - \rho\eta\dot{\theta} \qquad \text{(V-46)}$$

and

$$\rho\dot{\psi} \equiv \rho\dot{\varepsilon}-\rho\eta\dot{\theta}-\rho\theta\dot{\eta} \qquad \text{(V-47)}$$

The desired result, i.e. (V-44),now follows from the substitution of (V-46) and (V-47) into (V-45).

We now want to proceed with expressing (V-45) in a global form which will be valid for all points p in a thermoelastic body with symmetry uniformity. The real key to our work here is the assumption that the symmetry group of the heat flux function $\underset{\sim}{\ell}$ is the same as that of the response function $\underset{\sim}{S}$, where both are taken relative to some fixed elastic symmetry atlas $\underset{\sim}{\Lambda}$ on $\mathcal{B}$; therefore, relative to $\underset{\sim}{\Lambda}$, $\underset{\sim}{\ell}$ is a smooth global field on $\mathcal{B}$ with

$$\underset{\sim}{\ell}(\underset{\sim}{F},\theta,\underset{\sim}{g},p) = \underset{\sim}{\ell}_{\underset{\sim}{r}^{\alpha}(p)}(\underset{\sim}{F},\theta,\underset{\sim}{g})$$

for any elastic symmetry chart $(\mathcal{U}_{\alpha},\underset{\sim}{r}^{\alpha}(\cdot)) \in \underset{\sim}{\Lambda}$ such that $p\in\mathcal{U}_{\alpha}$.

Exercise 42 If we set $\ell^{i}_{jk}(\underset{\sim}{F},\theta,p) \equiv \partial\ell^{i}/\partial F^{jk}$, show that $\forall \underset{\sim}{G}\in G_{\underset{\sim}{\Lambda}}$

$$\ell^{i}_{jk}(\underset{\sim}{F}\underset{\sim}{G},\theta,g,p)G^{k}_{m} = \ell^{i}_{jm}(\underset{\sim}{F},\theta,\underset{\sim}{g},p), \tag{V-48}$$

$\forall(\underset{\sim}{F},\theta,\underset{\sim}{g})$, and that $\forall \underset{\sim}{K}\in \underset{\sim}{g}_{\underset{\sim}{\Lambda}}$

$$\ell^{i}_{jk}(F,\theta,\underset{\sim}{g},p)F^{jm}K^{k}_{m} = 0 \tag{V-49}$$

hint: use the fact that

$$\ell^{i}(\underset{\sim}{F}\underset{\sim}{G},\theta,\underset{\sim}{g},p) = \ell^{i}(\underset{\sim}{F},\theta,\underset{\sim}{g},p), \quad \forall \underset{\sim}{G}\in G_{\underset{\sim}{\Lambda}},$$

and all $(\underset{\sim}{F},\theta,p)$, and then proceed as in Chapter II, §9.

Once again, we denote the motion of $\mathcal{B}$, which induces global time-dependent coordinates x^{i} on $\mathcal{B}$, by $\phi(t)$. The heat flux in $\phi_t(\mathcal{B})$ is determined, of course, by

$$q^{i}(\phi_t(p)) = \ell^{i}(\underset{\sim}{F}_{p}, \theta(p), \underset{\sim}{g}(p), x^{m}(p)) \tag{V-50}$$

where $\underset{\sim}{F}_p = \phi_{*p}(t) \circ \underset{\sim}{r}^{\alpha}(p)^{-1}$. Substituting (V-50) into the equation of heat transfer (V-44) yields the local equation,

$$\ell^i_{jk} \frac{\partial F^{jk}}{\partial x^i} + \ell^i_\theta g_i + \ell^{ij}_{\underset{\approx}{g}} \frac{\partial g_j}{\partial x^i} + \ell + \rho r = \rho \theta \dot{\eta} \tag{V-51}$$

which is valid (only) on U_α; in (V-51) we have set

$$\ell^i_\theta = \frac{\partial \ell^i}{\partial \theta}, \quad \ell^{ij}_{\underset{\approx}{g}} = \frac{\partial \ell^i}{\partial g_j}, \quad \ell = \frac{\partial \ell^i}{\partial x^i}$$

<u>Exercise</u> 43 Prove that ℓ^i_θ, $\ell^{ij}_{\underset{\approx}{g}}$, and ℓ are all independent of the choice of the symmetry chart $(U_\alpha, \underset{\sim}{r}^\alpha(\cdot)) \in \underset{\sim}{\Lambda}$.

As a result of the above exercise we are again faced with the problem of converting only a single expression in our field equation, i.e. $\ell^i_{jk} \frac{\partial F^{jk}}{\partial x^i}$, from a local to a global form. In order to do this we again introduce an elastic symmetry connection H, whose connection symbols relative to (x^i) are functions $\Gamma^i_{jk}(x^m)$; then, as a consequence of (V-49), we obtain from (V-51) and the condition

$$[F^{-1i}_{\ j} (\frac{\partial F^i_k}{\partial x^m} + \Gamma^i_{lm} F^l_k)] \in \underset{\sim}{g}_\Lambda, \quad m = 1,2,3$$

the system of equations

$$- \ell^i_{jk} F^{nk} \Gamma^j_{ni} + \ell^i_\theta g_i + \ell^{ij}_{\underset{\approx}{g}} \frac{\partial g_j}{\partial x^i} + \ell + \rho r = \rho \theta \dot{\eta} \tag{V-52}$$

Exercise 44 Show that $\bar{\ell}_k^{ij} = \ell_{kn}^i(\underset{\sim}{F},\theta,\underset{\sim}{g},x^m)F^{jn}$ is a global field on $\mathcal{B}$ by using (V-48).

Exercise 45 If we again take $\kappa: \mathcal{B} \to R^3$ as a fixed reference configuration which induces a global coordinate system (X^A) on $\mathcal{B}$ show that, relative to κ, the system (V-52) assumes the (global) form

$$\tilde{\ell}_k^{iA}\left(\frac{\partial^2 x^k}{\partial X^A \partial X^B} - {}^{\kappa}\Gamma_{AB}^{C}\frac{\partial x^k}{\partial X^C}\right)\frac{\partial X^B}{\partial x^j} \tag{V-53}$$

$$+ \ell_\theta^i g_i + \ell_{\underset{\sim}{g}}^{ij}\frac{\partial g_j}{\partial x^i} + \ell + \rho r = \rho\theta\dot{\eta}$$

where $\tilde{\ell}_k^{iA} = \ell_{kl}^j(\underset{\sim}{F},\theta,\underset{\sim}{g},x^m)\overset{\kappa}{F}{}^{A\ell}$.

Exercise 46 In terms of the heat flux vector $\underset{\sim}{q}_\kappa$ taken relative to the reference configuration κ, the equation of heat transfer takes the form

$$\text{Div}\ \underset{\sim}{q}_\kappa + \rho_\kappa r = \rho_\kappa\theta\dot{\eta} \tag{V-54}$$

Show that if we define a heat flux function $\underset{\sim}{Q}$ relative to the elastic symmetry atlas $\underset{\sim}{\Lambda}$ by $\underset{\sim}{Q}(\underset{\sim}{F},\theta,\underset{\sim}{g},p) \equiv (\det \underset{\sim}{F})\underset{\sim}{F}^{-1}\,\underset{\sim}{\ell}(\underset{\sim}{F},\theta,\underset{\sim}{g},p)$, so that $\underset{\sim}{\ell} = \frac{1}{(\det \underset{\sim}{F})}\underset{\sim}{F}\underset{\sim}{Q}(\underset{\sim}{F},\theta,\underset{\sim}{g},p)$, then

$$\ell_{jk}^i = \frac{1}{(\det \underset{\sim}{F})}[F_m^i Q_{jk}^m + (\delta_j^i\delta_{km} - F_m^i F_{kj}^{-1})Q^m \tag{V-55}$$

where $Q_{jk}^i \equiv \partial Q^i/\partial F^{jk}$. Substitute (V-55) into (V-53) and show that we obtain as our new set of equations

$$\tilde{\ell}^{AB}_{k}\left(\frac{\partial^2 x^k}{\partial X^A \partial X^B} - \overset{\kappa}{\Gamma}{}^{C}_{BA}\frac{\partial x^k}{\partial X^C}\right) - q^{A}_{\kappa}\overset{\kappa}{\theta}{}^{B}_{AB} \tag{V-56}$$

$$+ J\ell^i_\theta g_i + J\ell^{ij}_{\underset{\sim}{g}}\frac{\partial g_j}{\partial x^i} + J\ell + \rho_\kappa r = \rho_\kappa \theta \dot{\eta}$$

where $\tilde{\ell}^{AB}_{k} = \dfrac{1}{(\det \underset{\sim}{\overset{\kappa}{F}})}\ell^r_{kl}(\underset{\sim}{F},\theta,\underset{\sim}{g},x^m)\overset{\kappa}{F}{}^{A}_{r}\overset{\kappa}{F}{}^{Bl}$

Note also that

$$J\ell^i_\theta g_i = \frac{\partial q^A_\kappa}{\partial \theta}\overset{\kappa}{g}_A, \quad J\ell^{ij}_{\underset{\sim}{g}}\frac{\partial g_j}{\partial x^i} = \frac{\partial q^A_\kappa}{\partial \overset{\kappa}{g}_B}\frac{\partial \overset{\kappa}{g}_B}{\partial X^A},$$

and $J\ell = \dfrac{\partial q^A_\kappa}{\partial X^A}$ where $\overset{\kappa}{g}_A = g_i x^i{}_{,A}$. Now, let us turn our attention to the expression which appears on the right-hand side of the equation of heat transfer (V-44). By virtue of the constitutive relation (V-3') and the restriction (V-9) we clearly have

$$\eta(\phi_t(p)) = H_{\underset{\sim}{r}^{\alpha}(p)}(\underset{\sim}{F},\theta,p), \forall p \in U_\alpha, \tag{V-54}$$

and two distinct possibilities arise. First of all suppose that $\underset{\sim}{G_\Lambda} = \underset{\sim}{G^*_\Lambda}$; then it is easy to see that we may define a smooth global field on $\mathcal{B}$ via

$$H_{\underset{\sim}{\Lambda}}(\underset{\sim}{F},\theta,p) \equiv H_{\underset{\sim}{r}^{\alpha}(p)}(\underset{\sim}{F},\theta,p) \tag{V-58}$$

where $(U_\alpha, \underset{\sim}{r}^{\alpha}(\cdot)) \in \underset{\sim}{\Lambda}$ is any elastic symmetry chart such that $p\epsilon U_\alpha$ [18]. If we drop the $\underset{\sim}{\Lambda}$ subscript on the response function H and set

$$H_{ij}(\underset{\sim}{F},\theta,p) \equiv \partial H/\partial F^{ij}, \quad H_\theta(\underset{\sim}{F},\theta,p) \equiv \partial H/\partial\theta$$

then we have, $\forall\underset{\sim}{G}\epsilon G_{\underset{\sim}{\Lambda}}$,

$$H_{ij}(\underset{\sim}{F}\underset{\sim}{G},\theta,p)G_k^j = H_{ik}(\underset{\sim}{F},\theta,p) \tag{V-59}$$

and

$$H_\theta(\underset{\sim}{F}\underset{\sim}{G},\theta,p) = H_\theta(\underset{\sim}{F},\theta,p) \tag{V-60}$$

for all $(\underset{\sim}{F},\theta,p)$.

Exercise 47 Verify the results contained in (V-59) and (V-60).

Now, if we differentiate the component form of (V-58) through with respect to t we easily obtain

$$\dot{\eta} = H_{jk}\dot{F}^{jk} + H_\theta\dot{\theta} \tag{V-61}$$

where the first expression on the right-hand side is a local one, valid only for points $p\epsilon U_\alpha$; in order to convert this expression into a global one, which will be valid for

[18] this follows, of course, from the fact that if $p\epsilon U_\alpha \cap U_\beta$, where $(U_\alpha, \underset{\sim}{r}^\alpha(\cdot))$, $(U_\beta, \underset{\sim}{r}^\alpha(\cdot))$ are both elastic symmetry charts in $\underset{\sim}{\Lambda}$, then the coordinate transformations $\underset{\sim}{G}_{\alpha\beta}$, from $\underset{\sim}{r}^\beta(\cdot)$ to $\underset{\sim}{\tilde{r}}^\alpha(\cdot)$ will be elements of $G_{\underset{\sim}{\Lambda}}^*$ which, by virtue of $\tilde{V}$-9, is the symmetry group of $H_{\underset{\sim}{\Lambda}}$. The case $G_{\underset{\sim}{\Lambda}} = G_{\underset{\sim}{\Lambda}}^*$ includes all solid bodies.

all points $p\varepsilon\mathcal{B}$, we need only note that

$$F^{ij} = \overset{\kappa}{F}{}^{Aj}\frac{\partial x^l}{\partial X^A}\frac{\partial v^i}{\partial x^l} \qquad \text{(V-62)}$$

Substitution from (V-62) into (V-61) then produces the equation

$$\dot{\eta} = \tilde{H}^A_{j}\frac{\partial x^l}{\partial X^A}\frac{\partial v^j}{\partial x^l} + H_\theta\dot{\theta} \qquad \text{(V-63)}$$

where $\tilde{H}^A_{j} = H_{jk}(\underset{\sim}{F},\theta,p)\overset{\kappa}{F}{}^{Ak}$ is now a global field on $\mathcal{B}$. Of course, it may be the case that $G_{\underset{\sim}{\Lambda}} \neq G^*_{\underset{\sim}{\Lambda}}$ so that (V-58) does not define a smooth global field of response functions on $\mathcal{B}$. Even in this case, however, we know that we may combine (V-11) and the restriction (V-9) so as to obtain

$$H_{\underset{\sim}{r}^\alpha(p)}(\underset{\sim}{F}\underset{\sim}{G},\theta,p) = H_{\underset{\sim}{r}^\alpha(p)}(\underset{\sim}{F},\theta,p) \qquad \text{(V-64)}$$

$$+ H_{\underset{\sim}{r}^\alpha(p)}(\underset{\sim}{G},\theta,p) - H_{\underset{\sim}{r}^\alpha(p)}(\underset{\sim}{1},\theta,p)$$

for all $p\varepsilon U_\alpha$ and all $\underset{\sim}{G}\varepsilon G_{\underset{\sim}{\Lambda}}$; if we again set $H_{\underset{\sim}{\Lambda}}(\underset{\sim}{F},\theta,p) = H_{\underset{\sim}{r}^\alpha(p)}(\underset{\sim}{F},\theta,p)$, for <u>any</u> $(U_\alpha,\underset{\sim}{r}^\alpha(\cdot)) \varepsilon \underset{\sim}{\Lambda}$ such that $p\varepsilon U_\alpha$, then differentiation of (V-64) with respect to $\underset{\sim}{F}$ yields the fact that the gradient of $H_{\underset{\sim}{\Lambda}}$ is a global field on $\mathcal{B}$, i.e.,

$$H_{ij}(\underset{\sim}{F},\theta,p) = H_{\underset{\sim}{r}^\alpha(p)ij}(\underset{\sim}{F},\theta,p) \qquad \text{(V-65)}$$

for <u>all</u> $(U_\alpha,\underset{\sim}{r}^\alpha(\cdot)) \varepsilon \underset{\sim}{\Lambda}$ such that $p\varepsilon U_\alpha$. We can now proceed exactly as we did in the case where $G_{\underset{\sim}{\Lambda}} = G^{*}_{\Lambda}$

Finally, we want to indicate how we would obtain the global equations of motion and the global equation of heat transfer for the case where $\mathcal{B}$ is a thermoanelastic body with uniform symmetry instead of a thermoelastic body with uniform symmetry; the procedure is essentially the same one which was followed in Chapter IV where we deduced the global field equations of motion for a smooth materially uniform anelastic body from those of a materially uniform elastic body. In other words, we replace the elastic symmetry atlas $\underset{\sim}{\Lambda} = \{(\mathcal{U}_\alpha, \underset{\sim}{r}^{\alpha}(\cdot)), \alpha\varepsilon I\}$ by its associated anelastic symmetry atlas $\underset{\sim}{\Lambda}(t) = \{(\mathcal{U}_\alpha, \underset{\sim}{r}^{\alpha}(\cdot,t)), \alpha\varepsilon I\}$ where $\underset{\sim}{r}^{\alpha}(\cdot,t)$ and $\underset{\sim}{r}^{\alpha}(\cdot)$ are related via (IV-39). We also replace the elastic symmetry connection H by its associated anelastic symmetry connection H^t and transform $F_\alpha(\cdot) \to \underset{\sim}{F}_\alpha(\cdot,t)$ where $\underset{\sim}{F}_\alpha(\cdot,t) = \underset{\sim}{F}_\alpha(\cdot) \circ \underset{\sim}{r}^{\alpha}(\cdot,t) \circ \underset{\sim}{\alpha}(\cdot,t) \circ \underset{\sim}{r}^{\alpha}(\cdot,t)^{-1}$ on $\mathcal{U}_\alpha$. Finally we replace the connection symbols $\Gamma^{A}_{BC}(\cdot)$ of H by the corresponding connection symbols $\Gamma^{A}_{BC}(\cdot,t)$ of H^t.

Exercise 46. Show that the equations of motion (V-36) can be rewritten in the form

$$\left(\tilde{H}^{iA}_{jk} x^{k}_{A,B} + \tilde{H}^{i}_{jB}\right) \frac{\partial X^B}{\partial x^i} + \tilde{H}^{i}_{j\theta} g_i + \rho b_j = \rho \ddot{x}_j \tag{V-66}$$

where $x^{k}_{A,B}$ denotes the covariant derivative of $x^{k}_{A} \equiv \frac{\partial x^k}{\partial X^A}$ relative to the connection H and

$$\tilde{H}^{iA}_{jk} = H^{iA}_{jk}(\underset{\sim}{x}_{*}\circ\underset{\sim}{F},\theta,X^K)F^A_B = \frac{\partial S^i_{j}(\underset{\sim}{F},\theta,X^K)}{\partial F^k_A} F^A_B \tag{V-67}$$

$$\tilde{H}^i_{jB} = H^i_{jB}(\underset{\sim}{x}_{*}\circ\underset{\sim}{F},\theta,X^K) = \frac{\partial S^i_{j}(\underset{\sim}{F},\theta,X^K)}{\partial X^B} \tag{V-68}$$

$$\tilde{H}^i_{j\theta} = H^i_{j\theta}(\underset{\sim}{x}_{*}\circ\underset{\sim}{F},\theta,X^K) = \frac{\partial S^i_{j}(\underset{\sim}{F},\theta,X^K)}{\partial \theta}$$

If we now make the indicated transformations in the field equations of motion (V-66) and take into account the definitions of the global fields $\tilde{H}^{iA}_{jk}$, $\tilde{H}^i_{jB}$, and $\tilde{H}^i_{j\theta}$ then it is a straightforward matter to show that these balance equations become

$$(\tilde{K}^{iA}_{jk}x^k_{A|B} + \tilde{K}^i_{jB})\frac{\partial X^B}{\partial x^i} + \tilde{K}^i_{j\theta}g_i + \rho b_j = \rho\ddot{x}_j \tag{V-70}$$

where

$$\tilde{K}^{iA}_{jk} = \tilde{H}^{iB}_{jk}(\underset{\sim}{x}_{*}\underset{\sim}{\alpha}\underset{\sim}{F},\theta,X^K)\alpha^A_C F^C_B \tag{V-71}$$

$$\tilde{K}^i_{jB} = \tilde{H}^i_{jB}(\underset{\sim}{x}_{*}\underset{\sim}{\alpha}\underset{\sim}{F},\theta,X^K) \tag{V-72}$$

$$\tilde{K}^i_{j\theta} = \tilde{H}^i_{j\theta}(\underset{\sim}{x}_{*}\underset{\sim}{\alpha}\underset{\sim}{F},\theta,X^K) \tag{V-73}$$

We recall that the argument $\underset{\sim}{x}_{*}\underset{\sim}{\alpha}\underset{\sim}{F}$ has components $x^i_A\alpha^A_B F^B_C$ and that $F^A_B(X^K,t) = \alpha^A_C(X^K,t)F^C_B(X^K)$. If we also make note of the relation

$$\alpha^A_C x^k_{A|B} = (\alpha^A_C x^k_A)_{,B}$$

between the covariant derivatives relative to the elastic symmetry connection H and the associated anelastic symmetry connection $\overset{t}{H}$ then the balance equations (V-70) can be rewritten in the form

$$\hat{K}^{iA}_{jk}(\alpha^C_A x^k_C)_{,B} + \tilde{K}^i_{jB})\, \frac{\partial X^B}{\partial x^i} \qquad \text{(V-74)}$$

$$+ \tilde{K}^i_{j\theta} g_i + \rho b_j = \rho \ddot{x}_j$$

In this case, the global field $\hat{K}^{iA}_{jk}$ is given by

$$\hat{K}^{iA}_{jk} = H^{iB}_{jk}(\underset{\sim}{x}_{*}\underset{\sim}{\alpha}\underset{\sim}{F},\theta,X^K)F^A_B$$

so that (V-74) is precisely the same as (V-66) except that the variable x^k_A in (V-66) has been replaced, throughout the equation, by the variable $x^k_B \alpha^B_A$.

<u>Exercise</u> 47 Derive the variant of the system (V-74) which is based on the response function for the Piola-Kirchhoff stress tensor.

Finally, we want to consider the global form which the equation of heat transfer (V-44) assumes in a thermoanelastic body; to accomplish this we first rewrite the global balance equation (V-53) in the equivalent form

$$(\tilde{\ell}^{iA}_{k} x^{k}_{A,B} + \tilde{\ell}^{i}_{B}) \frac{\partial X^{B}}{\partial x^{i}} + \tilde{\ell}^{i}_{\theta} g_{i} + \tilde{\ell}^{ij}_{\underset{\sim}{g}} \frac{\partial g_{j}}{\partial x^{i}} + \rho r = \rho\theta\dot{\eta} \qquad \text{(V-75)}$$

where

$$\tilde{\ell}^{iA}_{k} = \ell^{iB}_{k}(\underset{\sim}{x}_{*}\underset{\sim}{F}, \theta, \underset{\sim}{g}, X^{K}) F^{A}_{B} \qquad \text{(V-76)}$$

$$\tilde{\ell}^{i}_{B} = \ell^{i}_{B}(\underset{\sim}{x}_{*}\underset{\sim}{F}, \theta, \underset{\sim}{g}, X^{K}) \qquad \text{(V-77)}$$

$$\tilde{\ell}^{i}_{\theta} = \ell^{i}_{\theta}(\underset{\sim}{x}_{*}\underset{\sim}{F}, \theta, \underset{\sim}{g}, X^{K}) \qquad \text{(V-78)}$$

$$\tilde{\ell}^{ij}_{\underset{\sim}{g}} = \ell^{ij}_{\underset{\sim}{g}}(\underset{\sim}{x}_{*}\underset{\sim}{F}, \theta, \underset{\sim}{g}, X^{K}) \qquad \text{(V-79)}$$

and $\ell^{iA}_{k} = \dfrac{\partial \ell^{i}(F, \theta, g, X^{K})}{\partial F^{k}_{A}}$

$$\ell^{i}_{\theta} = \frac{\partial \ell^{i}(F, \theta, g, X^{K})}{\partial \theta}$$

$$\ell^{ij}_{\underset{\sim}{g}} = \frac{\partial \ell^{i}(F, \theta, g, X^{K})}{\partial g_{j}}$$

$$\ell^{i}_{A} = \frac{\partial \ell^{i}(F, \theta, g, X^{K})}{\partial X^{A}}$$

are the gradients of the response function $\underset{\sim}{\ell}$ relative to the elastic symmetry atlas $\underset{\sim}{\Lambda}$. Once again we carry out the transformations $\underset{\sim}{\Lambda} \rightarrow \underset{\sim}{\Lambda}(t)$ and $H \rightarrow H^{t}$ and in so doing we

$$(\tilde{m}^{iA}_{k} x^{k}_{A|B} + \tilde{m}^{i}_{B}) \frac{\partial X^{B}}{\partial x^{i}} + \tilde{m}^{i}_{\theta} g_{i} + \tilde{m}^{ij}_{\underset{\sim}{g}} \frac{\partial g_{j}}{\partial x^{i}} + \rho r = \rho\theta\dot{\eta} \qquad \text{(V-80)}$$

where $\tilde{m}^{iA}_{k}$, $\tilde{m}^{i}_{B}$, $\tilde{m}^{i}_{\theta}$, and $\tilde{m}^{ij}_{g}$ are global fields given by

$$\tilde{m}^{iA}_{k} = \ell^{iB}_{k}(\underset{\sim}{x}_{*}\underset{\sim}{\alpha}\underset{\sim}{F}, \theta, \underset{\sim}{g}, X^{K}) \alpha^{A}_{C} F^{C}_{B}$$

$$\tilde{m}^{i}_{B} = \ell^{i}_{B}(\underset{\sim}{x}_{*}\underset{\sim}{\alpha}\underset{\sim}{F}, \theta, \underset{\sim}{g}, X^{K})$$

$$\tilde{m}^{i}_{\theta} = \ell^{i}_{\theta}(\underset{\sim}{x}_{*}\underset{\sim}{\alpha}\underset{\sim}{F}, \theta, \underset{\sim}{g}, X^{K})$$

and

$$\tilde{m}^{ij}_{\underset{\sim}{g}} = \ell^{ij}_{\underset{\sim}{g}}(\underset{\sim}{x}_{*}\underset{\sim}{\alpha}\underset{\sim}{F}, \theta, \underset{\sim}{g}, X^{K})$$

If we now use the relationship between the covariant derivatives with respect to H and H^{t} then we can rewrite (V-80) as

$$(\tilde{n}^{iA}_{k}(\alpha^{C}_{A} x^{k}_{C})_{,B} + \tilde{m}^{i}_{B}) \frac{\partial X^{B}}{\partial x^{i}} + \tilde{m}^{i}_{\theta} g_{i} + \tilde{m}^{ij}_{\underset{\sim}{g}} \frac{\partial g_{j}}{\partial x^{i}} + \rho r = \rho\theta\dot{\eta} \qquad \text{(V-81)}$$

where the global field $\tilde{n}^{iA}_{k}$ is given by

$$\tilde{n}^{iA}_{k} = \ell^{iB}_{k}(\underset{\sim}{x}_{*}\underset{\sim}{\alpha}\underset{\sim}{F}, \theta, \underset{\sim}{g}, X^{K}) F^{A}_{B},$$

i.e., (V-81) is the same as (V-75) except that, once again, the variable x^{k}_{A} is replaced by the variable $x^{i}_{B}\alpha^{B}_{A}$ throughout the equation. We now rewrite the global expression for $\dot{\eta}$, i.e. (V-63), in the form

$$\dot{\eta} = \tilde{h}^{A}_{j} x^{i}_{A} \frac{\partial v^{j}}{\partial x^{i}} + \tilde{h}_{\theta}\dot{\theta} \tag{V-82}$$

where the global fields $\tilde{h}^{A}_{j}$ and $\tilde{h}_{\theta}$ are given by

$$\tilde{h}^{A}_{j} = h^{B}_{j}(\underset{\sim}{x}_{*}\underset{\sim}{F},\theta,X^{K})F^{A}_{B} \tag{V-83}$$

and

$$\tilde{h}_{\theta} = h_{\theta}(\underset{\sim}{x}_{*}\underset{\sim}{F},\theta,X^{K})$$

with

$$h^{B}_{j} = \frac{\partial H(F,\theta,X^{K})}{\partial F^{i}_{B}} , \quad h = \frac{\partial H(F,\theta,X^{K})}{\partial\theta}$$

We again replace $\underset{\sim}{\Lambda}$ by $\underset{\sim}{\Lambda}(t)$ and H by H^{t} and obtain

$$\dot{\eta} = \tilde{k}^{A}_{j} x^{i}_{B} \alpha^{B}_{A} \frac{\partial v^{j}}{\partial x^{i}} + \tilde{k}_{\theta}\dot{\theta} \tag{V-84}$$

where $\tilde{k}^{A}_{j}$ and $\tilde{k}_{\theta}$ are given by

$$\tilde{k}^{A}_{j} = h^{B}_{j}(\underset{\sim}{x}_{*}\underset{\sim}{\alpha}\underset{\sim}{F},\theta,X^{K})F^{A}_{B}$$

and

$$\tilde{k}_{\theta} = h_{\theta}(\underset{\sim}{x}_{*}\underset{\sim}{\alpha}\underset{\sim}{F},\theta,X^{k}).$$

Putting all our results together we have the following equation of heat transfer for a thermoanelastic body with uniform symmetry:

$$(\tilde{n}^{iA}_{k}(\alpha^{B}_{A}x^{k}_{B})_{,C} + \tilde{m}^{i}_{c}) \frac{\partial X^{C}}{\partial x^{i}} + \tilde{m}^{i}_{\theta} g_{i} + \tilde{m}^{ij}_{\underset{\sim}{g}} \frac{\partial g_{j}}{\partial x^{i}} + \rho r \tag{V-85}$$

$$= \rho\theta(\tilde{k}^{A}_{j}\alpha^{B}_{A}x^{i}_{B} \frac{\partial v^{j}}{\partial x^{i}} + \tilde{k}_{\theta}\dot{\theta}) + \rho r$$

Exercise 48 Derive the variant of (V-85) which obtains if we use the response function for the heat flux vector relative to the reference configuration κ.

Chapter VI Some Recent Directions in Current Research

We indicate here, just very briefly, some recent trends in research on anelasticity theory, dislocation motions, and relations with plasticity. In [39], Reinicke and Wang have taken up the study of flow rules associated with anelastic bodies of the type considered by Wang and Bloom [19]. Recall that the flow rule in the theory of anelasticity is the governing first-order differential equation $\dot{\underset{\sim}{\alpha}} = \underset{\sim}{\psi}$ where $\underset{\sim}{\alpha}(p,t)$ is the anelastic transformation function in some local motion $\underset{\sim}{r}(p,t)$;$\underset{\sim}{\alpha}(p,t)$ statisfies the "initial" condition $\underset{\sim}{\alpha}(p,t) \to id_{\mathcal{B}p}$ as $t \to \infty$. Some general results concerning the flow rule have already been delineated in Chapter IV (§4). In [39] a special type of flow rule which is appropriate for theories of elastic-plastic response is considered. With $\underset{\sim}{r}(p)$ a fixed local reference configuration, $\underset{\sim}{r}(p,t)$ a local motion, $A_{\underset{\sim}{r}(p)} = \underset{\sim}{r}(p) \circ \underset{\sim}{\alpha}(p,t) \circ \underset{\sim}{r}(p)^{-1}$ the relative anelastic transformation function, and $F_{\underset{\sim}{r}(p)}(t) \equiv \underset{\sim}{r}(p,t) \circ \underset{\sim}{r}(p)^{-1}$ the deformation gradient, Reinicke and Wang identify $A_{\underset{\sim}{r}(p)}(t)^{-1}$ as the plastic part of $F_{\underset{\sim}{r}(p)}(t)$ and $E_{\underset{\sim}{r}(p)}(t) \equiv F_{\underset{\sim}{r}(p)}(t) A_{\underset{\sim}{r}(p)}(t)$ as the elastic part of $F_{\underset{\sim}{r}(p)}(t)$. At any instant t in the local process $\underset{\sim}{r}(p,t)$ the plastic part of the deformation gradient, $A_{\underset{\sim}{r}(p)}(t)^{-1}$, is fixed and the relative anelastic response function $\underset{\sim}{S}^{t}_{\underset{\sim}{r}(p)}$ is just the elastic reponse function $S_{\underset{\sim}{r}(p)}$ evaluated at the elastic part $E_{\underset{\sim}{r}(p)}(t)$ for all $F_{\underset{\sim}{r}(p)}(t)$. [see IV-6 for the definition of $S^{t}_{\underset{\sim}{r}(p)}$

In materials with elastic-plastic response it is usually assumed that the plastic deformation remains constant in an unloading process and that the differential increment of the plastic deformation is a linear function of the differential increment of the total deformation in a loading process. A yield criterion is used to distinguish between loading and unloading processes, the yield criterion being defined by a yield surface. For an anelastic point p Reinicke and Wang define the yield criterion by a tensor field $\underset{\sim}{\lambda}(\underset{\sim}{r}(p,t),\underset{\sim}{\alpha}(p,t))$ whose values are considered to be linear functions on the rates of change $\dot{\underset{\sim}{r}}(p,t)$ of arbitrary local processes. Then $\dot{\underset{\sim}{r}}(p,t)$ is a loading direction if

$$\underset{\sim}{\lambda}(\underset{\sim}{r}(p,t),\underset{\sim}{\alpha}(p,t))[\dot{\underset{\sim}{r}}(p,t)] > 0 \qquad \text{(VI-1)}$$

and $\dot{\underset{\sim}{r}}(p,t)$ is an unloading direction direction if (VI-1) does not hold. A local configuration $\underset{\sim}{r}(p,t)$ at time t in a process is said to correspond to a point in the elastic range of the elastic-plastic material if

$$\underset{\sim}{\lambda}(\underset{\sim}{r}(p,t),\underset{\sim}{\alpha}(p,t)) = \underset{\sim}{0} \qquad \text{(VI-2)}$$

as in this case all increments from $\underset{\sim}{r}(p,t)$ are unloading increments. But if

$$\underset{\sim}{\lambda}(\underset{\sim}{r}(p,t),\underset{\sim}{\alpha}(p,t)) \neq \underset{\sim}{0}$$

then the subspace of all tensors $\underset{\sim}{\tau}$ for which $\underset{\sim}{\lambda}(\underset{\sim}{r}(p,t),\underset{\sim}{\alpha}(p,t))[\underset{\sim}{\tau}] = 0$ corresponds to the tangent plane of the yield surface as all directions $\underset{\sim}{r}(p,t)$ on the positive side of this subspace with $\lambda(\underset{\sim}{r}(p,t),\underset{\sim}{\alpha}(p,t))[\dot{\underset{\sim}{r}}(p,t)] > 0$ are loading directions while the subspace itself and the negaitve side of the subspace with $\underset{\sim}{\lambda}(\underset{\sim}{r}(p,t),\underset{\sim}{\alpha}(p,t))[\underset{\sim}{r}(p,t)] \leq 0$ are formed by unloading directions. The flow rule in an unloading process has the form

$$\dot{\underset{\sim}{\alpha}}(p,t) = 0 \qquad \text{(VI-4)}$$

so that $\underset{\sim}{\alpha}(p,t)$ remains constant in such a process while in a loading process the authors [39] choose as a flow rule

$$\dot{\underset{\sim}{\alpha}}(p,t) = \underset{\sim}{\Phi}(\underset{\sim}{r}(p,t),\underset{\sim}{\alpha}(p,t))[\dot{\underset{\sim}{r}}(p,t)] \qquad \text{(VI-5)}$$

where $\underset{\sim}{\Phi}$ is a fourth-order tensor field whose values are linear transformations on the rate of change $\dot{\underset{\sim}{r}}(p,t)$. Examples of processes $\underset{\sim}{r}(p,t)$ which contain both loading and unloading periods are considered as is the special case where $\underset{\sim}{\Phi}$ is a field in $\underset{\sim}{r}(p,t)$ only independent of $\underset{\sim}{\alpha}(p,t)$, the anelastic transformation function; except in this latter special case for which (VI-5) is integrable, a fixed functional relation between $\underset{\sim}{r}(p,t)$ and $\underset{\sim}{\alpha}(p,t)$ does not exist and the flow rule (VI-4),(VI-5) together with the "initial" condition $\underset{\sim}{\alpha}(p,t) \to id_{\mathcal{B}p}, t \to +\infty$, must be solved in order to determine the anelastic transformation function $\underset{\sim}{\alpha}(p,t)$ in each given process $\underset{\sim}{r}(p,t)$. Reinicke and Wang then proceed to analyse the restriction placed on the flow rule by material symmetry and material frame-indifference and conclude by showing that the local flow rule introduced above has a global representation on the whole body manifold $\mathcal{B}$.

In [40] Reinicke considers the propagation of accelaration waves in a materially uniform smooth anelastic body using the special flow rule formulated in [39]. As the flow rule itself distinguishes a loading process from an unloading process there result two distinct propagation conditions in [40] for acceleration waves in anelastic bodies, one for loading waves and one for unloading waves. The propagation conditions turn out to be generalizations of the usual propagation conditions for acceleration waves in elastic bodies with the growth and decay of the amplitudes of acceleration waves computed by use of the theory of characteristic surfaces and bicharacteristics. Reinicke [40] also derives transport equations for loading and unloading acceleration waves of multiplicity one and these equations, which govern the variation of the wave amplitudes along the bicharacteristics are generalizations of the usual transport equations for acceleration waves in an elastic body. In the course of the presentation in [40] the global equations of motion for anelastic bodies and bodies with elastic range are derived following the general presentation of Wang [42]; these derivations are formally equivalent to the one presented in Chapter IV (§8) for anelastic material bodies only. Finally we note that Reinicke [40] derives a differential equation of Bernoulli type, as the equation governing the amplitudes of acceleration waves propagating in anelastic material bodies, as opposed to the transport equation of Riccatti type obtained by Bloom [41] in his study of one-dimensional acceleration waves in anelastic materials; as Reinicke [40] indicates the difference in the two results is due to the assumption in [40] that $[\dot{\alpha}^A_B] \neq 0$ in a loading wave, an assumption which is absent in presentation given in [41].

Bibliography

(1) Kondo, K. "Geometry of Elastic Deformation and Incompatability", Memoirs of the Unifying Study of the Basic Problems in Engineering Sciences by Means of Geometry, Vol. I, Division C, Tokyo: Gakujutsu Benken Fukyu-Kai, 1955.

(2) Kondo, K., "Non-Riemannian Geometry of Imperfect Crystals from a Macroscopic Viewpoint", Memoirs of the Unifying study of the Basic Problems in Engineering Sciences by Means of Geometry, Vol. I, Division D, Tokyo: Gakujutsu Benken Fukyu-Kai, 1955.

(3) Bilby, B. A., Progress in Solid Mechanics 1 pp. 329-398 (1960) Ed. I. N. Sneddon & R. Hill.

(4) Kröner, E., "Allegemeine Kontinuumstheorie der Versetzungen und Eigenspannungen. Arch. Rat. Mech. Anal., 4, 273-334 (1960).

(5) Truesdell, C. and Noll, W., "The Non-Linear Field theories of Mechanics" §44, Encyclopedia of Physics, Vol. III/3, Springer, Berlin - Heidelberg - New York, 1965.

(6) Noll, W., "Materially Uniform Simple Bodies with Inhomogeneities", Arch. Rational Mech. Anal., Vol. 27, pp. 1-32, 1967.

(7) Wang, C. C., "On the Geometric Structures of Simple Bodies, a Mathematical Foundation for the Theory of Continuous Distributions of Dislocations", Arch., Rational Mech. Anal., Vol. 27, pp. 33-92, 1967.

(8) Toupin, R. A., "Dislocated and Oriented Media", Mechanics of Generalized Continua, E. Kröner, Ed., Springer-Verlag, Berlin, pp. 126-140, 1968.

(9) Bilby, B. A., "Geometry and Continuum Mechanics", Mechanics of Generalized Continua, E. Kröner, Ed., Springer-Verlag, Berlin, pp. 180-198, 1968.

(10) Kröner, E., "Interrelations between Various Branches of Continuum Mechanics", Mechanics of Generalized Continua, E. Kröner, Ed., Springer-Verlag, Berlin, pp. 330-340, 1968.

(11) Wang, C. C. and Bowen, R. M., "Acceleration Waves in Inhomogeneous Isotropic Elastic Bodies", Arch. Rational Mech. Anal. Vol. 38, pp. 13-44, 1971.

(12) Bowen, R. M. and C. C. Wang, "Thermodynamic Influences on Acceleration Waves in Inhomogeneous Isotropic Elastic Bodies with Internal State Variables", Arch. Rat. Mech. Anal., 41, 287- , (1971)

(13) Bowen, R. M. & C. C. Wang, "Acceleration Waves in Orthotropic Elastic Materials", Arch. Rat. Mech. Anal., 47, 149-170 (1972)

(14) Wang, C. C., "Universal Solutions for Incompressible Laminated Bodies", Arch. Rational Mech. Anal., Vol. 29, pp. 161-192, 1968.

(15) Eckart, C., "The Thermodynamics of Irreversible Processes, IV. The Theory of Elasticity and Anelasticity", Phys. Rev., Vol. 73, pp. 373-382, 1948.

(16) Truesdell, C., The Mechanical Foundations of Elasticity and Fluid Dynamics, Gordon and Breach, Science Publishers, Inc., N. Y. (1965).

(17) Truesdell, C. & Toupin, R., "The Classical Field Theories" Encyclopdia of Physics, Vol. III/1, Springer, Berlin-Heidelberg - New York, 1960.

(18) Bloom, F., Dislocation Motion and Materially Uniform Quasi-Elastic Bodies, Ph.D. Thesis, Cornell Univ., September 1971.

(19) Bloom, F. and C. C. Wang, "Material Uniformity and Inhomogeneity in Anelastic Bodies". Arch. Rational Mech. Anal. 53, 246-276, (1974).

(20) Bloom, F. and C. C. Wang, "Global Thermodynamic Field Equations of Balance for Anelastic Bodies". Arch. Mech. Stos. (in Press).

(21) Kobayashi, S. and Nomizu, K., Foundations of Differential Geometry, Vol. I., Interscience Publ., New York-London, 1963.

(22) Wang, C. C., Multilinear Algebra, Euclidean Spaces and Differential Manifolds, the Johns Hopkins Univ., Dept. of Mech., Balt, Md., 1965-66.

(23) Wang, C. C., Applications of Differential Geometry Parts II and Part III, The John Hopkins Univ., Dept. of Mech., Balt, Md., 1967-68.

(24) Truesdell, C. & C. C. Wang, Introduction to Rational Elasticity, Noordhoff International Publishing, Leyden (1973).

(25) Noll, W., Proc. Sym. Applied Math, Vol. XVII, 93-101 (1965).

(26) Noll, W., Toupin, R. A., and C. C. Wang, "Continuum Theory of Inhomogeneities in Simple Bodies", Springer-Verlag, Berlin, Heidelberg, New York, 1968.

(27) Trudesdell, C., Six Lectures on Natural Philosophy, Springer-Verlag, (1968).

(28) Noll, W., "On the Continuity of the Solid and Fluid States", Journal of Rational Mechanics and Analysis, 4, 3-81 (1955).

(29) Noll, W., "A Mathematical Theory of the Mechanical Behavior of Continuous Media", Arch. Rat. Mech. Anal., 2, 197-226 (1958).

(30) Nono, T., "On the Symmetry Groups of Simple Materials: An Application of the Theory of Lie Groups", J. Math. Anal. Appl., 24, 110-135, (1968).

(31) Belinfante, J., "Lie Algebras and Inhomogeneous Simple Materials", Siam J. Appl. Math. 25, 260-268 (1973).

(32) Wang, C. C., "Generalized Simple Bodies", Arch. Rational Mech. Anal., Vol. 32, pp. 1-28, 1969.

(33) Wang, C. C. and Bowen, R. M., "On the Thermodynamics of Non-Linear Materials with Quasielastic Response", Arch. Rational Mech. Anal., Vol. 22, 1966.

(34) Coleman, B. D., "Thermodynamics of Materials with Memory", Arch. Rat. Mech. Anal., 17, 1-46, (1964).

(35) Coleman, B. D., "On Thermodynamics Strain Impulses, and Viscoelasticity, Arch. Rat. Mech. Anal. 17, 230-254, (1964).

(36) Wang, C. C., "On the Thermodynamics of Inhomogeneous Bodies", in Fundamental Aspects of Dislocation Theory N. B. S. Special Publication 317 II, (1970).

(37) Coleman, B. D. and W. Noll, "The Thermodynamics of Elastic Materials with Heat Conduction and Viscosity", Arch. Rat. Mech. Anal., 13, 167-178 (1963).

(38) Truesdell, C., "A Theorem on the Isotropy Groups of a Hyperelastic Material", Proc. Nat. Acad. Sci. U. S. A, 52, 1081-1083, (1964).

(39) Reinicke, K. M. and C. C. Wang, "On the Flow Rules of Anelastic Bodies", Arch. Rat. Mech. Anal., Vol. 58, (1975), 103-113.

(40) Reinicke, K. M. "Acceleration Waves in Anelastic Material Bodies", Arch. Rat. Mech. Anal., Vol. 62 (1976), 165-187.

(41) Bloom, F., "Growth and Decay of One-Dimensional Acceleration Waves in Nonlinear Materials with Anelastic Response", Int. J. Engng. Sci., Vol. 13 (1975), 353-367.

(42) Wang, C. C., "Global Equations of Motion for Anelastic Bodies and Bodies with Elastic Range", Arch. Rat. Mech. Anal., Vol. 59 (1975), 1-23.